Advances in Intelligent Systems and Computing

Volume 1210

The series "Advances in Intelligent Systems and Computing" contains publications on theory, applications, and design methods of Intelligent Systems and Intelligent Computing. Virtually all disciplines such as engineering, natural sciences, computer and information science, ICT, economics, business, e-commerce, environment, healthcare, life science are covered. The list of topics spans all the areas of modern intelligent systems and computing such as: computational intelligence, soft computing including neural networks, fuzzy systems, evolutionary computing and the fusion of these paradigms, social intelligence, ambient intelligence, computational neuroscience, artificial life, virtual worlds and society, cognitive science and systems, Perception and Vision, DNA and immune based systems, self-organizing and adaptive systems, e-Learning and teaching, human-centered and human-centric computing, recommender systems, intelligent control, robotics and mechatronics including human-machine teaming, knowledge-based paradigms, learning paradigms, machine ethics, intelligent data analysis, knowledge management, intelligent agents, intelligent decision making and support, intelligent network security, trust management, interactive entertainment, Web intelligence and multimedia.

The publications within "Advances in Intelligent Systems and Computing" are primarily proceedings of important conferences, symposia and congresses. They cover significant recent developments in the field, both of a foundational and applicable character. An important characteristic feature of the series is the short publication time and world-wide distribution. This permits a rapid and broad dissemination of research results.

**** Indexing: The books of this series are submitted to ISI Proceedings, EI-Compendex, DBLP, SCOPUS, Google Scholar and Springerlink ****

More information about this series at http://www.springer.com/series/11156

Matteo Zallio
Editor

Advances in Human Factors in Robots, Drones and Unmanned Systems

Proceedings of the AHFE 2020 Virtual Conference on Human Factors in Robots, Drones and Unmanned Systems, July 16–20, 2020, USA

Springer

Editor
Matteo Zallio
Stanford University
Stanford, CA, USA

ISSN 2194-5357 ISSN 2194-5365 (electronic)
Advances in Intelligent Systems and Computing
ISBN 978-3-030-51757-1 ISBN 978-3-030-51758-8 (eBook)
https://doi.org/10.1007/978-3-030-51758-8

This Springer imprint is published by the registered company Springer Nature Switzerland AG
The registered company address is: Gewerbestrasse 11, 6330 Cham, Switzerland

Advances in Human Factors and Ergonomics 2020

AHFE 2020 Series Editors

Tareq Z. Ahram, Florida, USA
Waldemar Karwowski, Florida, USA

11th International Conference on Applied Human Factors and Ergonomics and the Affiliated Conferences

Proceedings of the AHFE 2020 Virtual Conference on Human Factors in Robots, Drones and Unmanned Systems, and the International Conference on Human Factors in Cybersecurity, July 16–20, 2020, USA

Advances in Neuroergonomics and Cognitive Engineering	Hasan Ayaz and Umer Asgher
Advances in Industrial Design	Giuseppe Di Bucchianico, Cliff Sungsoo Shin, Scott Shim, Shuichi Fukuda, Gianni Montagna and Cristina Carvalho
Advances in Ergonomics in Design	Francisco Rebelo and Marcelo Soares
Advances in Safety Management and Human Performance	Pedro M. Arezes and Ronald L. Boring
Advances in Human Factors and Ergonomics in Healthcare and Medical Devices	Jay Kalra and Nancy J. Lightner
Advances in Simulation and Digital Human Modeling	Daniel N Cassenti, Sofia Scataglini, Sudhakar L. Rajulu and Julia L. Wright
Advances in Human Factors and Systems Interaction	Isabel L. Nunes
Advances in the Human Side of Service Engineering	Jim Spohrer and Christine Leitner
Advances in Human Factors, Business Management and Leadership	Jussi Ilari Kantola, Salman Nazir and Vesa Salminen
Advances in Human Factors in Robots, Drones and Unmanned Systems	Matteo Zallio
Advances in Human Factors in Cybersecurity	Isabella Corradini, Enrico Nardelli and Tareq Ahram

(continued)

(continued)

Advances in Human Factors in Training, Education, and Learning Sciences	Salman Nazir, Tareq Ahram and Waldemar Karwowski
Advances in Human Aspects of Transportation	Neville Stanton
Advances in Artificial Intelligence, Software and Systems Engineering	Tareq Ahram
Advances in Human Factors in Architecture, Sustainable Urban Planning and Infrastructure	Jerzy Charytonowicz
Advances in Physical, Social & Occupational Ergonomics	Waldemar Karwowski, Ravindra S. Goonetilleke, Shuping Xiong, Richard H.M. Goossens and Atsuo Murata
Advances in Manufacturing, Production Management and Process Control	Beata Mrugalska, Stefan Trzcielinski, Waldemar Karwowski, Massimo Di Nicolantonio and Emilio Rossi
Advances in Usability, User Experience, Wearable and Assistive Technology	Tareq Ahram and Christianne Falcão
Advances in Creativity, Innovation, Entrepreneurship and Communication of Design	Evangelos Markopoulos, Ravindra S. Goonetilleke, Amic G. Ho and Yan Luximon

Preface

This book deals with an area of critical importance in both the digital society and in the field of human factors: "Robots, Drones and Unmanned Systems" Researchers are conducting cutting-edge investigations in the area of unmanned systems to inform and improve how humans interact with robotic platforms. Many of the efforts focused on refining the underlying algorithms that define system operation and on revolutionizing the design of human–system interfaces. The multi-faceted goals of this research are to improve ease of use, learnability, suitability, and human–system performance, which in turn will reduce the number of personnel hours and dedicated resources necessary to train, operate, and maintain the systems. As our dependence on unmanned systems grows along with the desire to reduce the manpower needed to operate them across both the military and commercial sectors, it becomes increasingly critical that system designs are safe, efficient, and effective. Optimizing human–robot interaction and reducing cognitive workload at the user interface require research emphasis to understand what information the operator requires, when they require it, and in what form it should be presented, so they can intervene and take control of unmanned platforms when it is required. With a reduction in manpower, each individual's role in system operation becomes even more important to the overall success of the mission or task at hand. Researchers are developing theories as well as prototype user interfaces to understand how best to support human–system interaction in complex operational environments. Because humans tend to be the most flexible and integral part of unmanned systems, the human factors and unmanned systems' focus considers the role of the human early in the design and development process in order to facilitate the design of effective human–system interaction and teaming.

This book gathers the proceedings of the AHFE 2020 Conference on Human Factors in Robots, Drones and Unmanned Systems, which was held on July 16–20, 2020, virtually. It addresses a variety of professionals, researchers, and students in the broad field of robotics, drones, and unmanned systems who are interested in the design of multi-sensory user interfaces (auditory, visual, and haptic), user-centered design, and task-function allocation when using artificial intelligence/automation to offset cognitive workload for the human operator. We hope it is informative, but even more so that it is thought-provoking. We hope it provides inspiration, leading the reader to formulate new, innovative research questions, applications, and potential solutions for creating effective human–system interaction and teaming with robots and unmanned systems.

Section 1 Human-AI-Robot Teaming
Section 2 Human Interaction with Manned and Unmanned Aerial Vehicles
Section 3 Novel Techniques of Human-Robot Interaction and Exoskeleton

Each section contains research papers that have been reviewed by members of the International Editorial Board. Our sincere thanks and appreciation to the board members as listed below:

July 2020 Matteo Zallio

Contents

Human-AI-Robot Teaming

Tasking, Teaming, Swarming: Design Patterns for Human Delegation of Unmanned Vehicles

Axel Schulte(✉), Felix Heilemann, Sebastian Lindner, and Diana Donath

Institute of Flight Systems (IFS), Universität der Bundeswehr Munich (UBM),
Werner-Heisenberg-Weg 39, 85577 Neubiberg, Germany
{axel.schulte,felix.heilemann,sebastian.lindner,
diana.donath}@unibw.de

Abstract. In this article we discuss the distinction between human delegation modes of single and multiple unmanned vehicles (UV). To describe the co-agency of humans and highly automated unmanned systems (i.e. human autonomy teaming, HAT) a design and description language for HAT design patterns will be used to describe and distinguish between control modes of tasking, teaming and swarming. Concerning tasking, we advocate for a scalable delegation concept. A distinction of multi-UV teaming and swarming will be made from a human user's perspective. Finally, we will briefly report on some preliminary results from human-in-the-loop experimentation in our fast-jet research simulator, involving the cockpit-based control of UVs and swarms.

Keywords: Human autonomy teaming · Task-based delegation · Teaming · Swarming · Delegation modes · Cognitive agents · Unmanned vehicles

1 Introduction and Background

Manned-unmanned Teaming (MUM-T) becomes increasingly important in future military missions. It describes the interoperability of manned and unmanned vehicles (UV) to pursue a common mission objective. Both, the manned and the unmanned vehicles need to be employed in the same confined spatial, temporal, and mission-related context. In MUM-T, the **unmanned platform(s)**, as well as its/their mission payloads will be **commanded by the manned asset(s)**. From this, the major challenge for MUM-T technologies arises, i.e. to **master the high work demands posed on the human user(s)** arising from the multi-platform mission management and task execution. Therefore, technical solutions for MUM-T have to encompass, amongst others:

(1) Dedicated human-machine interaction and interface concepts: These need to be considered to exert Meaningful Human Control over manned and particularly also a number of unmanned systems, i.e. the platforms, and their payloads. In addition to high-level control automation, this also requires an adequate concept of workshare and function allocation, as well as appropriate interaction design, to master the complexity

M. Zallio (Ed.): AHFE 2020, AISC 1210, pp. 3–9, 2021.
https://doi.org/10.1007/978-3-030-51758-8_1

effects that come along with it. Controllability and transparency requirements shall be taken into consideration as well.
(2) Collaborative mission planning and vehicle control algorithms: These represent the automation functions supporting task allocation amongst the participating vehicles, manned and unmanned, as well as the highly automated task performance. Those functions can either be centralized, usually in the manned command vehicle, or distributed over the participating platforms

2 Tasking, Teaming, and Swarming Design Patterns

In [1], we introduced a description method and a common language to structure and depict configurations for highly automated work systems involving humans (), cognitive agents (), and conventional automation (). Those actors can be attributed to different roles in the work system (i.e. Worker or Tool). The Worker takes initiative for the pursuit of the work process, whereas the Tool executes given commands. The relationships between the actors can either be hierarchical (i.e. delegation, tasking:), or heterarchical (i.e. assistance, teaming:). These elements describe Human Autonomy Teaming (HAT) design patterns and enable the construction of rather complex automated work systems. From thereon, systems engineering requirements for developing cognitive agents, and related human-agent interaction modalities are derived.

Figure 1 shows two elementary delegation design patterns for **tasking**, i.e. the delegation of tasks to a subordinate cognitive agent or a conventional tool. Sheridan's term of Human Supervisory Control and succeeding works on human-automation function allocation describe design options for the delegation of well defined, simple tasks to conventional tools, as depicted in Fig. 1a. With the advent of intelligent automation [2], also the delegation of complex tasks to a cognitive agent became feasible (see Fig. 1b).

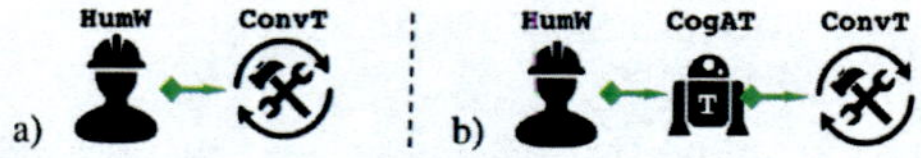

Fig. 1. Elementary design patterns for tasking (HumW: Human Worker, CogAT: Cognitive Agent as Tool, ConvT: Conventional Tool).

As opposed to pure tasking (or delegation) to a single subordinate agent or tool, **teaming** entails the introduction of some heterarchical structures ()amongst a number of actors in the work system, i.e. the team members (see Fig. 2).

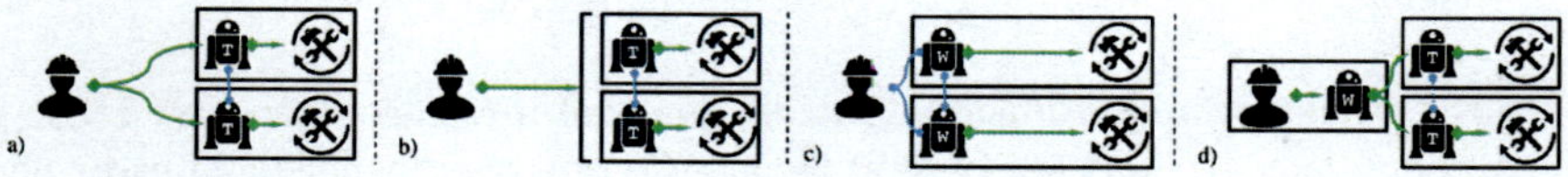

Fig. 2. Design patterns for teaming (T: Agent as Tool, W: Agent as Worker).

The metaphor of teaming stems from human-human organizational structures. We transfer it to socio-technical systems. This might involve purely unmanned teams, but

also the human user becoming teammate with artificial agents. In teaming, each member has a role to play, knows which roles are assigned to the other team members, and how they contribute to the overall work objective. Teaming is coordination at the task level, where each team member contributes their capabilities to complete the task.

Figure 2 shows four different design patterns for teaming. The patterns exemplify only two unmanned assets, for the sake of clarity and space. In the applications we tested, we implemented manned/unmanned teams of 3 to 5 UVs plus one command vehicle.

a) In this design option, the user delegates **individual tasks** to a small number of agents, each of which controls its given conventional tool(s) (e.g. UVs) [3]. The box indicates co-location (i.e. agent installed aboard the UV). Most of the coordination work in this pattern is done by the human user. However, there will also be local coordination of tasks amongst the dislocated UV-agents. This pattern can be regarded as a weak form of teaming, tasking is the dominating mode here.
b) As opposed to pattern a), here a more **complex team task** will be issued to all agents (indicated by bracket). Teaming amongst the dislocated agents will be facilitated by pursuing co-operative goals, and coordination amongst each other. Meitinger [4] investigated this design option by implementing cognitive agents on board of up to **five UVs, dynamically negotiating the distribution of sub-tasks. Gangl** [5] integrated this solution in a fast-jet cockpit simulator and conducted human-in-the-loop experiments. Generally, the chosen high level of autonomy and the complexity of teaming behaviors were found to be compromising situation awareness, controllability, and also causing complacency effects.
c) In this option, the cognitive agents, still co-located with their given tools, adopt the role of a Worker (i.e. pursue the given work objective by own initiative) [6]. Together with the human, they will form a **cooperating human-agent team**. Meitinger [7] investigated a solution, where a human pilot and four cooperating agents, each aboard a UV, performed an aerial mission. This pattern can be regarded as the highest level of autonomy, as well as the strongest form of human-agent teaming. On the downside, human-in-the-loop experiments showed that complex emergent behaviors may appear only little transparent to the human user [7].
d) Finally, in this option, the human delegation, in principle, can be very similar to options a) and b). However, in this case the tasks will be passed to a central planning and coordination agent, which would usually be co-located with the user (e.g. the pilot aboard a command vehicle in MUM-T). The coordination work, in this case, can be shared and traded between the human user and the central coordination agent in a wide range. Heilemann [8] investigates the approach of **scalable delegation** on a conceptual and experimental level.

The notion of **swarming** stems from biology, describing moving in or forming a large or dense group of small, rather simple animals. The observed complexity of swarming behavior emerges from frequent and parallel, but usually simple and local interactions of the swarm members, based upon the exchange and adaptation of behavioral parameters (e.g. direction and speed of motion). Usually, all swarm members follow the same purpose. In our context, we borrow the term swarming to describe the co-action of a larger number (i.e. maybe greater ten) of technical vehicles that all serve a human user-provided purpose. Clough [9] describes a swarm as a collection of autonomous individuals that rely on local sensory perception and reactive behavior producing global behavior by the interaction results.

Applying the swarming metaphor to a technical application, the swarm needs to be tasked by a human user. Starting from the elementary patterns of tasking (cf. Fig. 1), we derive a pattern involving a **swarm avatar**, as shown in Fig. 3a. However, the direct tasking or interaction of the human with the many swarm members, according to the pattern depicted in Fig. 1a is not an option. It could be proven by Coppin [10] that swarm performance breaks down or significantly decreases when direct human interventions in swarming algorithms are allowed. The purpose of the swarm avatar is to provide a tasking interface to the human user, to exert meaningful control over the swarm. Therefore, the avatar has to translate the purpose of the user into parameters of the swarming algorithms. The avatar will usually be a tool agent, however, most likely co-located with the human command station. This concept also allows integrating a swarm into any teaming context, as illustrated in Fig. 3b. This teaming mode corresponds with the one in Fig. 2a. However, it is imaginable to substitute a swarm within any other teaming structure shown in Fig. 2 by using the avatar principle.

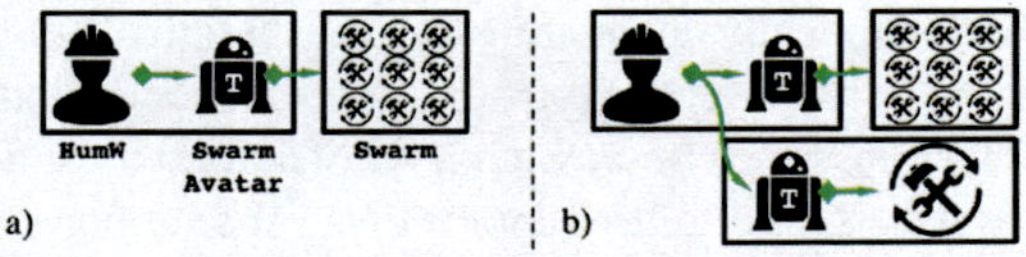

Fig. 3. Design pattern for tasking swarm through swarm avatar, and swarm as a team member.

3 Scalable Delegation

Based on the design pattern comprising the central planning and coordination agent (cf. Fig. 2d), we developed a scalable delegation approach, with which the user can interact on the level of individual tasks up to the level of team tasks [11]. The swarm avatar concept enables a consistent task delegation to the swarms in the same way as to the UVs. We suggest that this delegation process takes place on a combined timeline and task delegation interface, shown in Fig. 4, at the following scalable delegation levels.

(1) Team (Fig. 4a): The mission planner determines the best-suited team member(s) for the task(s) and inserts the task(s) at the best position in the plan. This delegation level is available for different types of tasks (individual or team). In case of a team task, the mission planner supplements the corresponding sub-tasks and allocates them to the team members [12].

(2) Individual: Here, an individual task is assigned to a specific team member. Individual delegation can take place at the following levels:

1. **Due Vehicle** (Fig. 4b): The user specifies the UV to perform the task. The mission planner determines the best position of the task in the task list of the selected UV.

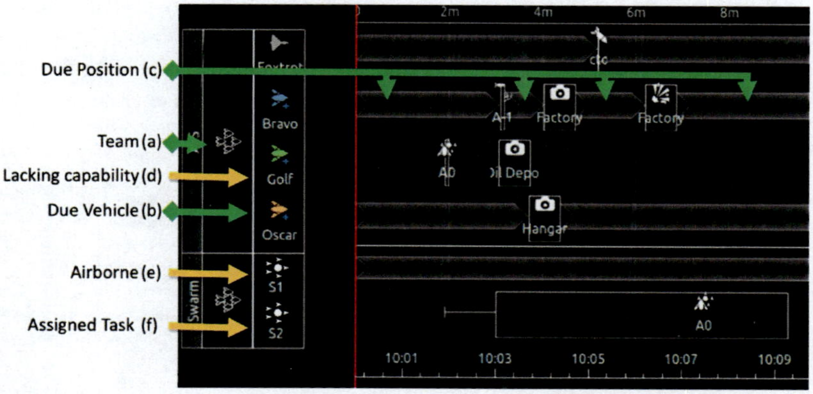

Fig. 4. Timeline-based task delegation interface, with scalable delegation levels.

2. **Due Position** (Fig. 4c): The user specifies the relative position of the delegated task in a vehicle's current task list. The mission planner then adjusts the timing of the dependent tasks and generates a new plan, according to the specified order of tasks.
3. **Due Time:** The user specifies the exact timing of the task, either by use of a parameter page or directly by inserting the task into the timeline and dragging it to the time it should take place.

The delegation interface provides direct feedback on the impact of each planning step on the resulting mission plan. Team members that do not display a delegation slot (such as UV 'Golf' in Fig. 4d), lack the capability required for the task (e.g. missing sensor). In the swarm section (Fig. 4e), all currently present swarms are shown. A new line appears if a new swarm has been launched. If a task has been assigned to the swarm (Fig. 4f), in addition to the task duration, the travel time from the launch point is shown.

4 Experimental Evaluation

For proof of concept, we implemented our multi-vehicle tasking, teaming and swarming system as a lab prototype, and integrated the human-automation interaction and mission planning functions in our fast-jet research flight simulator. Participants were eight German Air Force pilots, who had to perform six full MUM-T flight missions each, using one manned fighter aircraft and up to three unmanned combat aircraft and one air-launched reconnaissance swarm.

Figure 5a shows the frequency of use of the different delegation levels during the missions. The use of team tasks resulted in 105 individual tasks being given to the UVs by team delegation. For the delegation of individual tasks, the 'Due Position' delegation mode was preferred. Figure 5b shows that the delegation level strongly reflects the individual preference of each pilot (different colors). Figure 5c depicts some selected scores of a post-mission questionnaire. The scores show a great level of overall acceptance of our scalable tasking concept. The interaction concept was rated to be intuitive. The planning agent contributed to reach the mission goal faster. The pilots also considered the tasking concept as suitable for operational use in purely manned crews. Integrating UVs by means of teaming, led to a high appreciation of transparency and trust. The same holds true for our swarming approaches. Using the proposed concepts, we created a MUM-T system with which pilots could well imagine to operate in the future.

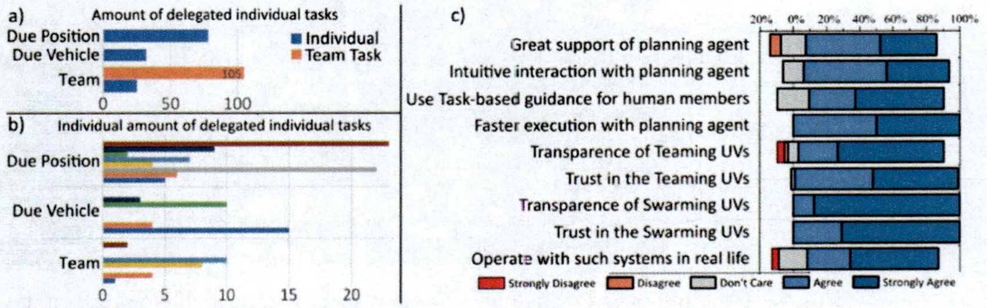

Fig. 5. Experimental results: a) usage distribution of delegated team and individual tasks, b) participants' delegations of the individual tasks, c) questionnaire results

5 Conclusion

To briefly sum up, the management of high work demands posed to the human user as well as the requisite transparency of delegation interfaces are amongst the major challenges. Scalable delegation modes of single, and even more UVs, provide an approach to increase the span of control while keeping work demands within limits. More into depth experimental data analyses are in preparation.

References

1. Schulte, A., Donath, D.: A design and description method for human-autonomy teaming systems. In: Karwowski, W., Ahram, T. (eds.) Intelligent Human Systems Integration. Advances in Intelligent Systems and Computing, vol. 722. Springer, Cham (2018)
2. Miller, C.A., Parasuraman, R.: Designing for flexible interaction between humans and automation: delegation interfaces for supervisory control. Hum. Factors **49**(1), 57–75 (2007)
3. Uhrmann, J., Schulte, A.: Concept, design, evaluation of cognitive task-based UAV guidance. Int. J. Adv. Intell. Syst. **5**(1 and 2), 145–158 (2012)
4. Schulte, A., Meitinger, C.: Introducing cognitive and co-operative automation into UAV guidance work systems. In: Human-Robot Interaction in Future Military Operations. Series Human Factors in Defence, Ashgate, pp. 145–170 (2010)
5. Gangl, S., Lettl, B., Schulte, A.: Management of multiple unmanned combat aerial vehicles from a single-seat fighter cockpit in manned-unmanned fighter missions. In: AIAA Infotech@Aerospace, p. 4899 (2013)
6. Onken, R., Schulte, A.: System-Ergonomic Design of Cognitive Automation – Dual-Mode Cognitive Design of Vehicle Guidance and Control Work Systems. Studies in Computational Intelligence, vol. 235. Springer, Heidelberg (2010)
7. Meitinger, C., Schulte, A.: Human-UAV co-operation based on artificial cognition. In: Harris, D. (ed.) Engineering Psychology and Cognitive Ergonomics, vol. 5639, pp. 91–100. Springer, Heidelberg (2009)

8. Heilemann, F., Schulte, A.: Time line based tasking concept for MUM-T mission planning with multiple delegation levels. In: Ahram, T., Karwowski, W., Vergnano, A., Leali, F., Taiar, R. (eds.) Advances in Intelligent Systems and Computing, vol. 1131. Springer, Cham (2020)
9. Clough, B.: UAV swarming? So what are those swarms, what are the implications, and how do we handle them? AUVSI unmanned systems. Technical report (2002)
10. Coppin, G., Legras, F.: Autonomy spectrum and performance perception issues in swarm supervisory control. Proc. IEEE **100**(3), 590–603 (2012)
11. Lindner, S., Schwerd, S., Schulte, A.: Defining generic tasks to guide UAVs in a MUM-T aerial combat environment. In: Karwowski, W., Ahram, T. (eds.) Advances in Intelligent Systems and Computing, vol. 903. Springer, Heidelberg (2019)
12. Heilemann, F., Schmitt, F., Schulte, A.: Mixed-initiative mission planning of multiple UCAVs from aboard a single seat fighter aircraft. In: AIAA Science and Technology Forum and Exposition (SciTech), p. 2205 (2019)

An Interaction Taxonomy of Human–Agent Teaming in Next Generation Combat Vehicle Systems

Craig J. Johnson[1](✉), Glenn J. Lematta[1], Lixiao Huang[3], Eric Holder[2], Shawaiz Bhatti[1], and Nancy J. Cooke[1]

[1] Arizona State University, Mesa, USA
{Cjjohn42,Glematta,Sabatt1,Nancy.cooke}@asu.edu
[2] CCDC Army Research Lab, Sierra Vista, USA
Eric.w.holder4.civ@mail.mil
[3] Center for Human, Artificial Intelligence, and Robot Teaming, Arizona State University, Mesa, USA
Lixiao.huang@asu.edu

Abstract. Next Generation Combat Vehicles (NGCVs) are incorporating more advanced technology which will enable humans and intelligent artificial agents to team up on the battlefield. Effective system design and evaluation for these human–agent teams require an understanding of individual and team tasks in the context of larger-scale operations. Previous taxonomies of human–automation interaction and human–agent teaming have been proposed, however, there is a lack of work focused on team interactions in the military domain and the teamwork dynamics required for our purposes are not captured. Unstructured interviews with subject matter experts, manuals, and relevant literature were synthesized, and a task analysis was conducted to develop a novel interaction taxonomy approach consisting of three primary categories, each with multiple dimensions: *task*, *team composition*, and *communication*. This taxonomy may generalize into human–agent teaming within a variety of NGCV crews and serve as a model for characterizing human–agent team interactions in other domains.

Keywords: Human–agent teaming · Taxonomy · Communication · Military

1 Introduction

Recent advances in autonomous technology suggest that near-future military warfighting systems will see more machine intelligence coordinating with humans [1]. These systems will be capable of executing complex and safety-critical tasks, such as maneuvering combat vehicles on a battlefield. As these individual system components are typically organized into teams and multi-team systems, the need for embedded autonomous agents to effectively coordinate and 'team up' with humans grows.

Armored vehicles such as the M1 Abrams Tank and M2 Bradley Fighting Vehicle have been demonstrated to be an effective means of providing mobile, protected

M. Zallio (Ed.): AHFE 2020, AISC 1210, pp. 10–16, 2021.
https://doi.org/10.1007/978-3-030-51758-8_2

firepower on the battlefield. However, recent initiatives such as the DoD *Unmanned Systems Integrated Roadmap FY2014–2042* [2] lay the groundwork for the integration of unmanned vehicles within the future of the US military. The unmanned variants of the Next Generation Combat Vehicles (NGCVs) show promise for fulfilling roles that have traditionally been filled exclusively by manned vehicles. These unmanned vehicles will have autonomous capabilities and will need to team up with dismounted Soldiers and manned vehicle crews. System interdependence means that these new systems will likely change the nature of traditional tasks and may introduce unintended consequences [3–5].

The term *agent* in this work refers to artificial agents (i.e., software agents or physical robots), whereas *human–agent* teaming refers to teams composed of at least one or more humans and one or more artificial agents. The focus of this work is on an approach to characterizing interactions in a specific configuration of NGCVs that includes two Robotic Combat Vehicles (RCVs), a Manned Combat Vehicle (MCV), and a human crew operating as a human–agent team.

1.1 Human–Agent Teaming

The active integration of heterogeneous perspectives is required to coordinate efforts and achieve collective goals on the battlefield. Team-level cognitive processes such as planning, reasoning, decision–making, and acting (i.e., team cognition) require team interactions [6]. Effective human–human teaming requires team player qualities such as predictability, directability, and observability in for communicating status information, making intentions known, and directing as well as being directed by others. These are challenges for human–agent teaming [5]. More research on team interactions and coordination is needed to support effective human–agent teaming design for applications such as NGCVs. The current work organizes and examines these types of interactions to support future work in human–agent teaming.

1.2 Interaction Taxonomies

We reviewed taxonomies developed in the areas of human–automation interaction, human–robot interaction, and human–agent teaming. Key concepts outlined in these taxonomies include classifying the levels of autonomy [7], robot types, interactions, and behaviors [8], core cognitive functions and human–agent teaming principles [9] and the composition, capacities, and interactions of the human–robot group [10]. Although these taxonomies provide a foundation for understanding, they do not integrate some important aspects of human–agent teaming interactions in military operations. A more detailed, domain-specific taxonomy that includes an understanding of the work context and possible interactions within NGCV crews is needed to assist in the development and evaluation of these and other future systems.

The taxonomy approach in this work is unique because it focuses on the interactions of the team, while also incorporating aspects of the team's composition and domain-specific tasks. The results of this study show potential to be generalized across a wider range of military human–agent teams and may serve as a model for characterizing interactions in human–agent teams in other domains.

2 Methods

The following procedures were employed to develop the taxonomy: (a) application scoping, (b) subject matter expert (SME) interviews, (c) military documentation review, (d) HAT academic literature review, (e) reflection on the first author's operational experience as a tank platoon leader, and (f) a task analysis. Only the scoping activities and SME interviews are discussed in detail here. Detail on the others can be found in [11].

An NGCV section consists of three vehicles—one Manned Combat Vehicle (MCV) and two remotely operated Robotic Combat Vehicles (RCVs)—and seven human crew members. All seven crew members are physically located in the MCV. Two operate the MCV, two dyads remotely operate the RCVs, and the seventh person functions as the vehicle commander and section leader providing higher-level coordination (see Fig. 1). Two sections form a platoon. The future RCV is anticipated to have a range of capabilities that will enable it to maneuver in various environments and under degraded conditions. Insight from existing research suggests many areas where the RCV may be more capable than its human crew members (e.g. sensing) or less capable (e.g. ambiguous situations) [4, 5].

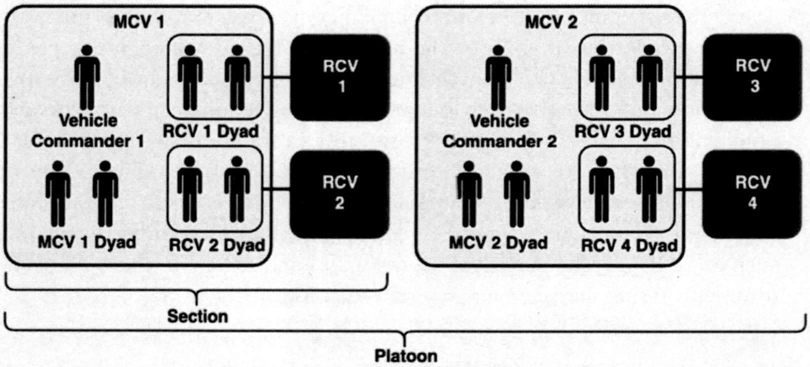

Fig. 1. Structure of a next generation combat vehicle (NGCV) platoon concept composed of two sections. There are seven members in each section who are physically located in a manned combat vehicle (MCV). Two dyads operate each RCV, one operates the MCV, and the vehicle commander provides higher-level coordination.

The scenario we examined is called Movement to Contact. It is a collective military operation that takes place when the tactical situation is unclear. It can be characterized by receipt of the mission from higher headquarters, preparation and dissemination of the mission within the team, a tactical movement to the objective, actions on contact, reconsolidation of forces following contact, and transition to new mission. Movement to Contact does not necessarily follow a linear progression. It contains a large number of tasks which can be generalized to other aspects of armored combat operations [12, 13].

SME interviews were conducted with two Army Officers with extensive first-hand experience in armored vehicle tactics. These interviews identified key interactions and challenges in existing operations such as the importance of communicating commander's intent to subordinates to allow adaptation to unanticipated situations, information filtering between subordinates and leaders, the desire to pull information from teammates when needed. Other key interactions included a reliance on shared understanding and implicit communication among experienced teams [11].

3 Results and Discussion

The output of the methodological approach was the development of a taxonomy structured for human–agent teaming interactions and applied to the Movement to Contact scenario for the MCV, RCVs and crew of an NGCV section. See [11] for a more detailed, earlier version of this taxonomy with examples. This taxonomy includes interactions within the human–agent team (human–human, human–agent, and agent–agent) as well as team-level interactions. Team interactions take many forms in sociotechnical environments. In a vehicle crew, information could be conveyed on a shared display, radio channel, or in physical space. These interactions can have different properties, such as meanings (e.g., status update, request), mechanisms (e.g., verbal behavior, touchscreen displays, auditory alerts) and communication flows [14, 15]. These properties are what make up the many of the elements of the taxonomy.

The taxonomy is split into three broad categories including *task*, *team composition*, and *communication*. Each category contains dimensions which can be used to characterize interactions (Table 1). The taxonomy is intended to capture elements important for the current context as part of an iterative process [11]. It should not necessarily be considered completely comprehensive and there may be some overlap between elements.

Table 1. Human–agent teaming interaction taxonomy categories and dimensions.

Categories	Task	Team composition	Communication
Dimensions	Task class	Entity types	Media
	Tasks	Roles	Mechanism
	Subtasks	Skill differentiation	Modality
	Essential interactions	Authority differentiation	Flow
		Temporal stability	Spatial proximity
		Interdependence	Temporal synchrony
			Content

3.1 Task Dimensions

Through the task analysis of RCV operators during movement to contact, we identified and categorized the tasks into three levels of abstraction: task classes, tasks, and subtasks. See [11] for the full task analysis.

Task Classes and Tasks. Task classes include *mobility*, *gunnery*, *actions on contact*, and *crew management*. 23 tasks were identified.

Subtasks. Subtasks are defined as lower-level tasks that are typically required to complete the higher-level tasks under nominal conditions. For example, the *detect surroundings for potential targets* task includes sub-tasks to operate weapons, designate sectors of responsibility, and communicate to operate weapons, etc. 169 total subtasks were identified.

Essential Interactions. Twenty-eight essential interactions spanning the entire task analysis were identified. Essential interactions were defined as interactions which (a) require interaction between two or more entities (humans or artificial) and (b) are essential to accomplish the higher-level task under nominal conditions. Failure to complete an essential interaction would be likely to result in negative outcomes.

3.2 Team Composition Dimensions

The team composition dimensions characterize the team members, their roles, and their relationships within the team.

Entity Types. Entity types characterize the members who make up a team or are involved in a specific interaction, distinguishing between humans and various types of artificial agents.

Roles. Roles describe the obligations, expectations, and responsibilities associated with a team member. Roles can be formal (e.g., Vehicle Commander), describing the formal title, position and/or responsibilities held, or task-dependent (e.g. sensor, decision-maker, effector).

Skill Differentiation. Skill differentiation characterizes how unique or shared skills are within the team [16, 17]. Teams may vary from highly differentiated to homogenous.

Authority Differentiation. Authority differentiation describes the degree to which authority is distributed among several individuals or centralized to a few or a single individual [16, 17].

Temporal Stability. Temporal stability describes the typical or expected lifecycle of a team [16, 17]. Teams may be intact for extended periods or only exist for a short time.

Interdependence. Interdependence describes the complementary relationships that two or more entities rely on to manage dependencies. Dependencies can be required (i.e. hard) when the capacity for a task is not possessed by one entity independently, or opportunistic (i.e., soft), where collaboration may improve efficiency or performance [5].

3.3 Communication Dimensions

Communication dimensions characterize the elements contained in communication between team members.

Media. Media describes the device, technology, or artifact that is utilized to interact. An example would be a vehicle intercom system [14]. Media considerations describe the actual tools that are implemented to support interactions.

Mechanism. Mechanism describes ways to meet communication needs (i.e., interactions) that are enabled by a medium such as text chat messages (a mechanism) displayed on a computer screen (media). A single medium may be able to support a variety of different interaction mechanisms such as graphical map features and text messages on a single computer screen [14]. Conversely, similar mechanisms (e.g., voice commands) may be involved in different media.

Modality. Modality describes the cognitive resource requirements for interaction such as "auditory" modality for spoken word over a radio It also includes the code (spatial or verbal) and processing stage (perception, cognition, responding) [18].

Communication Flow. Flow describes who communicates to who, defined by the sender and receiver. For example, when an RCV operator talks to the vehicle commander to share some information. Flow may include one–to–one communication or other combinations involving multiple entities [15, 19].

Spatial Proximity. Spatial proximity characterizes the physical distance between agents in an interaction. For instance, they can be either in the same place (i.e., co-located) or in different places (i.e., remote) [20].

Temporal Synchrony. Temporal synchrony characterizes how closely an individual interaction occurs in time. Interaction between agents can unfold at the same time or at different times (i.e., asynchronously) [20].

Communication Content. Communication content describes the information, syntax, and meaning contained within an interaction.

4 Conclusion

This work describes the development of a taxonomy for human–agent teaming interactions in a specific instantiation of NGCVs. This interaction taxonomy may help to guide the design and evaluation of human–agent teaming in these and other future military human–agent teams and has potential to be expanded into other domains such as robot-assisted search and rescue. Future work should include interactions between teams and across scales (e.g., platoons, companies, etc.) and seek to further refine, expand, and validate the taxonomy elements.

Acknowledgments. We thank our sponsor, the U.S. Army Research Laboratory, under the Cooperative Agreement No. W911-NF-18-2-0271. The views and conclusions contained in this document are those of the authors and should not be interpreted as representing the official policies, either expressed or implied, of the Army Research Laboratory or the U.S. government.

References

1. Cummings, M.: Artificial intelligence and the future of warfare. Chatham House for the Royal Institute of International Affairs (2017)
2. Fahey, K.M., Miller, M.J.: DoD Unmanned Systems Integrated Roadmap 2017–2042. Department of Defense. Technical report (2017)
3. Strauch, B.: Ironies of automation: still unresolved after all these years. IEEE Trans. Hum.-Mach. Syst. **48**(5), 419–433 (2017)
4. Endsley, M.R.: From here to autonomy: lessons learned from human–automation research. Hum. Factors **59**(1), 5–27 (2017)
5. Johnson, M., Bradshaw, J.M., Feltovich, P.J., Jonker, C.M., Van Riemsdijk, M.B., Sierhuis, M.: Coactive design: designing support for interdependence in joint activity. J. Hum.-Robot Interact. **3**(1), 43–69 (2014)
6. Cooke, N.J., Gorman, J.C., Myers, C.W., Duran, J.L.: Interactive team cognition. Cogn. Sci. **37**(2), 255–285 (2013)
7. Beer, J.M., Fisk, A.D., Rogers, W.A.: Toward a framework for levels of robot autonomy in human-robot interaction. J. Hum.-Robot Interact. **3**(2), 74–99 (2014)
8. Dudek, G., Jenkin, M., Milios, E.: A taxonomy of multirobot systems, Robot Teams Divers. Polymorph 3–22 (2002)
9. Save, L., Feuerberg, B., Avia, E.: Designing human-automation interaction: a new level of automation taxonomy. Proc Hum. Factors Syst. Technol (2012)
10. Yanco, H.A., Drury, J.: Classifying human-robot interaction: an updated taxonomy. In: 2004 IEEE International Conference on Systems, Man and Cybernetics (IEEE Cat. No. 04CH37583), vol. 3, pp. 2841–2846 (2004)
11. Huang, L., Johnson, C.J., Holder, E., Lematta, G.J., Bhatti, S., Barnes, M.J., Cooke, N.J.: Human–Autonomy Teaming: Interaction Metrics and Models for Next Generation Combat Vehicle Concepts (in press)
12. Mitchell, D.K.: Workload analysis of the crew of the Abrams V2 SEP: Phase I baseline IMPRINT model. Army Research Lab Aberdeen Proving Ground (2009)
13. US Army: ATP 3-20.15 Tank platoon. Headquarters, Department of the Army (2019)
14. Hollan, J., Stornetta, S.: Beyond being there. In: Proceedings of the SIGCHI Conference on Human Factors in Computing Systems, pp. 119–125 (1992)
15. Cooke, N.J., Gorman, J.C.: Interaction-based measures of cognitive systems. J. Cogn. Eng. Decis. Mak. **3**(1), 27–46 (2009)
16. Hollenbeck, J.R., Beersma, B., Schouten, M.E.: Beyond team types and taxonomies: a dimensional scaling conceptualization for team description. Acad. Manag. Rev. **37**(1), 82–106 (2012)
17. Baker, A.L., Schaefer, K.E., Hill, S.G.: Teamwork and Communication Methods and Metrics for Human–Autonomy Teaming. CCDC Army Research Laboratory Aberdeen Proving Ground (2019)
18. Wickens, C.D.: Multiple resources and mental workload. Hum. Factors **50**(3), 449–455 (2008)
19. Cooke, N.J., Gorman, J.C., Kiekel, P.A.: Communication as team-level cognitive processing. In: Macrocognition in Teams, pp. 51–64. CRC Press (2017)
20. Grudin, J.: Computer-supported cooperative work: history and focus. Computer **27**(5), 19–26 (1994)

Human-Autonomy Teaming for Unmanned Vehicle Control: Examination of the Role of Operator Individual Differences

Elizabeth Frost[1](✉), Heath Ruff[2], and Gloria Calhoun[1]

[1] Air Force Research Laboratory, 711 HPW/RHWC, Dayton, OH, USA
{Elizabeth.Frost.7,Gloria.Calhoun}@us.af.mil
[2] Infoscitex, Dayton, OH, USA
Heath.Ruff.ctr@us.af.mil

Abstract. Autonomous capabilities are reaching a point where they can fulfill the role of a teammate for the command and control of unmanned vehicles. Individual characteristics of a human operator may influence how an autonomous teammate is utilized and the team's performance. Twenty-four participants completed a questionnaire that included the Ten-Item Personality Inventory (TIPI), Desirability of Control Scale, and items regarding video game experience and propensity to trust. They then worked with either a human or autonomous teammate to complete a series of missions using multiple simulated unmanned vehicles and rated how much they trusted their teammate. Results found several correlations between TIPI scores and performance measures. Propensity to trust scores were correlated with their trust ratings when the teammate was human, but not correlated when the teammate was autonomous. There were no significant correlations associated with video game experience or Desirability of Control. Implications of the results are discussed.

Keywords: Human-autonomy team · Unmanned vehicles · Individual differences

1 Introduction

New and enhanced autonomous capabilities are enabling more advanced human-autonomy teaming. In the command and control (C2) domain, this is leading to examining humans and autonomous teammates working together to control unmanned vehicles (UVs). For instance, in what ways do individual characteristics of a human operator influence how an autonomous teammate is utilized and the team's overall performance?

Teaming with autonomy is relatively new and brings unique challenges that need to be addressed. One such challenge is overcoming the potential of an inherent bias towards not trusting an autonomous teammate as much as another human [1]. Having an accurate level of trust is crucial in human-autonomy teams because, if not appropriately calibrated, it can lead to underutilization or over-reliance on automation [2]. Being able to measure

M. Zallio (Ed.): AHFE 2020, AISC 1210, pp. 17–23, 2021.
https://doi.org/10.1007/978-3-030-51758-8_3

an individual's propensity to trust may help identify when someone is more prone to an inappropriate level of trust in the teammate which can degrade team performance [3].

Team Performance. Human-autonomy teams also bring new advantages such as being able to leverage the increased experience that individuals have with other advanced technological systems. Therefore, as C2 systems evolve, those with experience using complex systems, such as video games, may improve team performance. Video game experience has been linked with greater skills in a variety of related areas such as enhanced perceptual and attentional abilities [4] and more efficient cognitive task switching [5]. With advanced C2 systems that have some similarities to complex environments experienced in strategic video games and that require many of the skills shown to be enhanced by video game experience, there is a potential that gamers will be able to learn these C2 system quicker and their C2 performance will be better than peers who are not gamers.

Additionally, the "Big Five" personality traits (Extraversion, Emotional Stability, Agreeableness, Openness to Experience, and Conscientiousness) [6] have been shown to be potential predictors of performance. A meta-analysis by Barrick and Mount (1991) reviewed the Big Five in relation to various job performance criteria and across different occupational groups [7] and found that Conscientiousness was consistently related to all performance criteria. However, Extraversion was a strong predictor for managers and jobs that involved social interaction. This indicates that a member's role in a team could influence which personality traits are relevant for predicting performance.

Team Structure. The structure of the team determines each member's roles and responsibilities, which can have an effect on the team's performance [8]. For example, with an 'operator-driven' team structure, the main operator fulfills a management role and is responsible for task delegation. On the other hand, with a 'role-driven' structure, members are relatively equal in terms of authority and are assigned tasks based on their predefined roles. An operator-driven team structure raises additional questions in terms of how the operator decides to share the workload with the teammate. This could include individual differences, such as desire for control, which can be related to a more controlling supervisory approach [9].

This study explored both human-human and human-autonomy teams using both the operator-driven and role-driven team structures. Individual differences, including video game experience, propensity to trust, personality, and desirability of control, were considered in terms of their correlation with the teams' overall task performance in UV C2 simulated missions.

2 Method

2.1 Participants

Twenty-four volunteers (19 males, 5 females) from local universities and the general population participated in this study. Ages ranged from 19 to 36 (M = 26.50, SD = 4.70). All participants reported normal or corrected normal vision and color vision.

2.2 Design

A 2 (Team Composition; between-subjects) x 2 (Team Structure; within-subject) mixed design was used. For Team Composition, participants were paired with either a human or autonomous teammate (both fulfilled by a confederate) to complete missions using two Team Structures: Operator-driven and Role-driven (counterbalanced across participants). Details of the design, method, and performance data results are reported in [10].

2.3 Simulation Testbed

The testbed consisted of two C2 stations (one for the participant and one for the teammate). Each C2 station ran the 'IMPACT' simulation, a prototype control station that integrated several autonomy advancements to support the C2 of multiple heterogeneous (air, ground, and sea) UVs during a base defense mission. More details are available in [11].

2.4 Base Defense Mission

Participants completed two base defense missions with each Team Structure. Each mission was 30-mintues long and included a series of tasks that were completed by allocating one or more UVs. The team was responsible for the same geographic area, however, each teammate had control of six different UVs (two air, two ground, and two sea). Vehicle control was preassigned and could not be changed during the mission.

Tasks were associated with base defense, ranging from answering commander queries to ensuring the best UV assets were relocated in response to breaches (e.g., fence alarms, escorts, and mortar fire). Each task was broken down into one or more sub-tasks. The tasks could be completed by the participant, the teammate, or by both team members working together (i.e., each teammate completing at least one sub-task). Initial task assignment was based on the Team Structure such that when using the Operator-driven structure, participants received all tasks and determined which to complete themselves and which to assign to their teammate. When using the Role-driven structure, tasks were automatically assigned to each teammate based on pre-defined roles. For both Team Structures, the participant and teammate could trade tasks as needed.

2.5 Procedure

Upon arrival, participants read the informed consent document and completed a pre-questionnaire that included demographics for age, gender, vision, and video game experience including level of experience (no experience to expert) and the average hours per week they spent playing within the past year. Participants also completed the Ten-Item Personality Inventory (TIPI) [12], the Desirability of Control Scale [13], and rated six items designed to measure their propensity to trust a teammate (with no indication of the teammate being either human or autonomous) using a 5-point Likert scale.

Participants were then trained on the IMPACT simulation testbed including the interfaces, controls, and interactions with their assigned teammate. Before proceeding, participants had to successfully complete a capstone mission. The participants and their assigned teammate then completed two missions using their first assigned Team Structure. After each mission, participants completed a questionnaire that included rating how much they trusted their 'teammate' (regardless of Team Composition). Participants then completed two missions using the other Team Structure, followed by the same post-mission questionnaires. Lastly, participants completed a final questionnaire, comparing the two team structures, and were debriefed.

2.6 Measures

Individual scores were calculated for each of the pre-questionnaire measures. Video game experience was based on the raw Likert scale rating. The TIPI and Desirability of Control Scale scores were calculated using the scoring methods described in the referenced publications. A single score for propensity to trust was calculated using the average rating, with a higher score indicating a greater tendency to trust a teammate.

Team performance, taking into consideration both the participant's and teammate's contributions, was measured objectively using the number of tasks completed, the time to complete the tasks (response time), and an overall mission performance score based on the priority of the tasks completed. The percentages of sub-tasks completed by participants, teammates, or uncompleted were also recorded. All team performance measures were collapsed across the two missions using the same Team Structure.

3 Results

For all data sets analyzed, outliers of raw data points were identified (data points outside 1.5 times the interquartile range) [14] and replaced by the grand mean. Approximately 3.6% of the data were identified as outliers. Pairs of variables were compared using bivariate correlation procedures by computing Pearson's correlation coefficient.

There were no significant correlations between self-reported video game experience (level of experience or hours per week) and mission performance. There were also no significant correlations between Desirability of Control scores and the percentage of sub-tasks completed by participants or the number tasks they assigned to their teammate.

The TIPI produced a score of Extraversion (E), Emotional Stability (ES), Agreeableness (A), Openness to Experience (O), and Conscientiousness (C) for each participant. Table 1 provides correlations between TIPI scores and the objective performance measures (overall and for each Team Structure). Participants scoring higher on Extraversion tended to have higher mission performance scores, complete more tasks, and have faster task response times. These correlations were also significant when considering only the Operator-driven Team Structure. Additionally, with the Operator-driven structure, a higher Extraversion score was associated with a higher percentage of subtasks completed by the participant and fewer uncompleted sub-tasks. Greater Emotional Stability was also correlated with more tasks completed and a higher mission performance score overall, as well as for the Role-driven structure. For both Team Structures, and overall, participants scoring higher on Emotional Stability completed a higher percentage of sub-tasks and had a lower percentage of uncompleted sub-tasks. The more agreeable participants were, the higher the percentage of sub-tasks completed by the teammate, both overall and with the Role-driven structure. Conversely, participants more open to experiences tended to have their teammates complete fewer sub-tasks. This is accompanied by the tendency of those more open to experiences to assign fewer tasks to the teammate with the Operator-driven structure ($r(22) = -.451$, $p = .027$). Conscientiousness did not correlate to any of the team performance measures.

Table 1. Correlations between TIPI scores and team performance measures.

	E	ES	A	O	C
Mission performance score	.491*	.490*	.004	−.120	.069
Operator-driven	.564**	.395	.015	−.264	.028
Role-driven	.251	.521**	−.016	.140	.118
Task completion (#)	.523**	.458*	.084	−.171	−.011
Operator-driven	.673***	.377	.157	−.149	.138
Role-driven	.021	.351	−.073	−.119	−.238
Task response time (s)	−.497*	−.378	−.093	.107	.052
Operator-driven	−.476*	−.296	−.235	.191	.027
Role-driven	−.325	−.389	.249	−.118	.082
Participant sub-tasks (%)	.346	.522**	−.309	.074	−.001
Operator-driven	.441*	.447*	−.308	.165	.096
Role-driven	.032	.421*	−.176	.032	−.167
Teammate sub-tasks (%)	−.013	−.061	.507*	−.450*	−.057
Operator-driven	−.015	−.087	.385	−.469*	−.122
Role-driven	.005	.084	.661***	−.080	.237
Uncompleted sub-tasks (%)	−.390	−.539**	−.010	.221	.118
Operator-driven	−.519**	−.440*	−.077	.208	−.021
Role-driven	.065	−.445*	.124	.129	.324

All r(22), * $p < .05$, ** $p < .01$, *** $p < .001$

Table 2 provides correlations between Propensity to Trust scores and Trust in Teammate ratings. When averaged across human and autonomous Team Compositions (labeled "Overall"), propensity to trust was significantly correlated with participants' ratings of trust in their teammates for the Role-driven Team Structure. However, when examining Team Compositions separately, propensity to trust was significantly correlated to the participants' trust ratings for both Team Structures as well as overall when the teammate was another human. For the participants working with an autonomous teammate, there were no significant correlations between their propensity to trust scores and their trust in teammate ratings.

Table 2. Correlations between Propensity to Trust scores and Trust in Teammate ratings.

	Propensity to trust		
	Overall $r(22)$	Human $r(10)$	Autonomous $r(10)$
Trust in teammate	.397	.618*	.270
Operator-driven	.238	.602*	.167
Role-driven	.465*	.597*	.345

* p < .05

4 Discussion

These findings identify several personality measures correlated with team performance. The positive correlation of Extraversion with team performance measures when using the Operator-driven structure is consistent with previous findings for management tasks [15]. Emotional Stability has also previously been found to be an important predictor of performance [16], which is in line with the positive correlation with overall mission performance reported herein. Additionally, higher Emotional Stability being associated with participants completing more tasks and having fewer incomplete tasks may reflect participants' ability to handle high mission demands without becoming overwhelmed.

The personality traits Openness to Experience and Agreeableness were not related to performance but instead may have influenced how task workload was shared. Those higher in openness to experience tended to complete more tasks themselves which may be due to a desire to be more involved [17]. On the other hand, participants who were more agreeable perhaps didn't care about being as hands-on, leading to accepting the predefined task assignments which by default led to more sub-tasks completed by the teammate.

In regards to propensity to trust a teammate, the consistent correlations with post-experiment trust ratings for when the teammate was human, in contrast to the lack of correlation when the teammate was autonomous suggests that this measure fails to serve as a predictor for human-autonomy teams and alternative indicators are needed. Note, however, the propensity to trust ratings were completed before participants were aware of whether they were working with a human or autonomous teammate – therefore participants may have assumed their teammate would be human and their propensity to trust scores are not indicative of biases against automation.

The lack of Desirability of Control findings may reflect that task sharing across the team was required to successfully complete the missions. Previous research has shown that individuals will relinquish control of specific tasks or actions, if it allows them to have greater control of the overall situation or outcome [18].

In summary, several individual differences were identified that were correlated with team performance and trust for both human-human and human-autonomy teams responsible for UV C2. Although many of these and other individual differences have been previously examined in a variety of human-human teams, the implications of these differences still need to be further explored in the context of human-autonomy teams.

Acknowledgment. This work was funded by the Air Force Research Laboratory.

References

1. Madhavan, P., Wiegmann, D.A.: Similarities and differences between human-human and human-automation trust: an integrative review. Theor. Issues Erg. Sci. **8**(4), 277–301 (2007)
2. Hoffman, R.R., Johnson, M., Bradshaw, J.M., Underbrink, A.: Trust in automation. IEEE Intell. Syst. **28**(1), 84–88 (2013)
3. Ferguson, A.J., Peterson, R.S.: Sinking slowly: diversity in propensity to trust predicts downward trust spirals in small groups. J. Appl. Psychol. **100**(4), 1012 (2015)
4. Spence, I., Feng, J.: Video games and spatial cognition. Rev. Gen. Psychol. **14**, 92–104 (2010)
5. Cain, M.S., Landau, A.N., Shimamura, A.P.: Action video game experience reduces the cost of switching tasks. Atten. Percept. Psychophys. **74**(4), 641–647 (2012)
6. Norman, W.T.: Toward an adequate taxonomy of personality attributes: replicated factor structure in peer nomination personality ratings. J. Abnorm. Soc. Psychol. **66**, 574 (1963)
7. Barrick, M.R., Mount, M.K.: The big five personality dimensions and job performance: a meta-analysis. Person. Psychol. **44**(1), 1–26 (1991)
8. Urban, J.M., Bowers, C.A., Monday, S.D., Morgan Jr., B.B.: Workload, team structure, and communication in team performance. Mil. Psychol. **7**(2), 123–139 (1995)
9. Legrain, P., Paquet, Y., D'Arripe-Longueville, F., Antonini Philippe, R.: Influence of desirability for control on instructional interactions and intrinsic motivation in a sport peer tutoring setting. Int. J. Sport Psychol. **42**(1), 69–83 (2011)
10. Frost, E.: Human-Autonomy Teaming: The Effects of Team Structure (Doctoral Dissertation), Wright State University (2019)
11. Draper, M., Rowe, A., Douglass, S., Calhoun, G., Spriggs, S., Kingston, D., Frost, E.: Realizing Autonomy via Intelligent Hybrid Control: Adaptable Autonomy for Achieving UxV RSTA Team Decision Superiority (No. AFRL-RH-WP-TR-2018-0005). 711 Human Performance Wing, Wright-Patterson Air Force Base, United States. Technical report (2018)
12. Gosling, S.D., Rentfrow, P.J., Swann, W.B.: A very brief measure of the big five personality domains. J. Res. Pers. **37**, 504–528 (2003)
13. Burger, J.M., Cooper, H.M.: The desirability of control. Motiv. Emot. **3**(4), 381–393 (1979)
14. Tukey, J.W.: Exploratory Data Analysis. Addison-Wesley, Boston (1977)
15. Calhoun, G., Ruff, H., Murray, C.: Multi-Unmanned Vehicle Supervisory Control: An Initial Evaluation of Personality Drivers. Infotech@ Aerospace, AIAA-2012-2527, pp. 1–22 (2012)
16. Judge, T.A., Erez, A.: Interaction and intersection: the constellation of emotional stability and extraversion in predicting performance. Person. Psychol. **60**(3), 573–596 (2007)
17. Rothmann, S., Coetzer, E.P.: The big five personality dimensions and job performance. SA J. Ind. Psychol. **29**(1), 68–74 (2003)
18. Burger, J.M., McWard, J., LaTorre, D.: Relinquishing control over aversive stimuli. In: Annual Meeting of the Western Psychological Association, Seattle, WA (1986)

The New Science of Autonomous Human-Machine Teams (a-HMT): Interdependence Theory Scales

W. F. Lawless(✉)

Paine College, 1235 15th Street, Augusta, GA 30902, USA
w.lawless@icloud.com

Abstract. Autonomous submarines. Drone wingmen. Hypersonic missiles. The evolution of autonomous human-machine teams (A-HMT) is occurring when the rapidity of making decisions have become central to military defense, operating complex systems, transportation, etc. Social science, however, offers little guidance for the science of A-HMTs. The problem with social science is its basis in rational methodological individualism (MI), likely at the root of its replication crisis and its inability to make predictions. MI has impeded the generalization of every theory that has used it, e.g., game theory, additive aggregation economics, assembling automata, political science and philosophy. In the laboratory and field, MI's rational collective theory tellingly fails in the presence of conflict, where interdependence theory thrives. Recently, however, social science has experimentally reestablished the value of interdependence to human team science, especially for the best of science teams, but not theoretically, making the results important but ad hoc. By rejecting MI in favor of interdependence theory, a phenomenon difficult to control in the laboratory, we have hypothesized for teams, found and replicated that the optimum size of a team minimizes its member redundancy. With interdependence theory, we have also found that, proportional to the complexity of the barriers faced by a team to completing its mission, intelligence is critical to a team's maximum entropy production (MEP); that whereas physical training promotes physical skills and whereas book knowledge promotes cognitive skills, these two skill sets are orthogonal to each other, resolving a long-standing experimental and theoretical conundrum; and, lastly, that the best determinations of social reality, decisions by a team, and decisions for the welfare of a society are based on the interdependence of orthogonal effects: the social harmonic oscillation of information driven by orthogonal pro-con poles, alternatively presenting one argument before an audience of neutral judges countered by its opposing argument. From this foundation, unlike traditional models based on MI, interdependence theory scales to integrate wide swaths of field evidence, e.g., bacteria gene and business mergers seeking MEP, but, if failing, leading to collapse (weak entropy production, WEP). Instead of predictions which fail in interdependent situations, the way forward for autonomous systems is to limit autonomy with checks and balances, similar to how free humans limit autonomy.

Keywords: Interdependence · Human-Machine Teams · Autonomy

M. Zallio (Ed.): AHFE 2020, AISC 1210, pp. 24–30, 2021.
https://doi.org/10.1007/978-3-030-51758-8_4

1 Introduction

At the moment when social science has an opportunity to contribute to the advancement of a physical science, the science of human-machine teams, the discipline has faltered with little of substance to offer. In this brief conceptual review, we attribute the cause of failure of social science to methodological individualism's (MI) rejection of social effects (interdependence) in favor of a rational theory of behavior; we briefly review game theory, Von Neumann's automata, and rational choice and consensus theory; we counter with the criticism that MI is unable to determine context [1]. Then we address our findings that support interdependence theory, including redundancy, tradeoff uncertainty, orthogonal relationships and social harmonic oscillators. We note similarities between interdependence and quantum theory. From the perspective of interdependence theory, we close by addressing Schrödinger's [2] question, "*What is life*?"

Failures occurred in social science, not only with questionnaires like implicit racial bias [3], but also with political predictions; e.g., Tetlock & Gardiner's [4] technique to seek consensus among the best forecasters, known as superforecasters, but they predicted incorrectly that Brexit would not occur and Trump would not become President.

2 Methodological Individualism. Rational Choice Theory

Arrow (p. 1, in [5]) contradicted MI by attributing the production, possession, and nature of knowledge to social effects:

> ... technical information in the economy is an especially significant case of an irreducibly social category in the explanatory apparatus of economics ...

Arrow concluded that methodological individualism (MI), exemplified by game theory (p. 4, in [5]), could not explain the acquisition of knowledge; e.g., in game theory, scientists establish context by determining a game's rules and payoffs before beginning an experiment with humans.

For Axelrod (p. 7–8, in [6]), games led to his two results: first, "the pursuit of self-interest by each [participant] leads to a poor outcome for all." And, second, this situation can be avoided when sufficient punishment exists to discourage competition. An example of the use of collective punishment in China comes from Friedman [7]:

> ... with the leak over the weekend of government documents describing ... a broad Chinese assault ... underway for several years on the ethnic minority Uighur community in ... Xinjiang ... [its] massive detention camps for "retraining" purposes and the separation of families on a scale that is startling even for China. Beijing clearly wants to break the back of Islam in the province.

Einstein's theory of relativity, Heisenberg's uncertainty principle, and Schrodinger's quantum mechanics recognize that physical "reality is not as it appears to us" [8]; and yet social scientists, including rational economists, persist in determining social reality based on individual perceptions that are unreliable, commonly leading to invalid concepts even about individuals' beliefs about their own actions [9].

Economic decisions are supposedly based on the rational choices that consistently maximize an individual's self-interest. But it has long been known that individual preferences in game theory do not predict the choices made in games [10]. Rational choice theorists correct for this mistake by replacing stated individual preferences with *imputed* preferences based on the observed behavior of individuals [11].

Other limits to rational choice theory have been determined, including biases, but since these produce consistent behavior, they have been incorporated (e.g., [12]).

Rational choice theory, the traditional model of decision-making, attempts to improve individual decisions by making them more consistent and in line with their preferences [11]. This model is popular in military circles, where it is known as the perception-action cycle. Israel [13] added the claim that mathematics provides the rational criterion for truth. Consistency, however, is not only prized by consensus-seeking rational choice theorists, but also by authoritarians and gang leaders [3].

Rational choice theory is based on three assumptions. First, it assumes that an individual's behaviors converge to what its brain sees and choses, where reality is sufficiently determinable to make individual decisions consistent. Second, trained observers (e.g., scientists), by observing consistent behavior, can impute the choices an individual makes, overcoming the inability of subjective self-reports to determine individual preferences. Third, if the information collected converges, it forms a consensus for a collective.

Critique: The primary requirement to operate in reality requires a consensus in the assessment of a situation. Absent a consensus about a context, social reality cannot be predetermined by individuals *de novo*. Rational choice theorists also need consistency to determine behavior; and they need a consensus among themselves to determine a collective's interpretation of a context. Mann [14], however, found that conflict disables rational choice. What is ignored by rational choice theorists is that social effects automatically arise as individuals immediately begin to agree and disagree with each other in every social setting as they strive to determine the context. In contrast, conflict is of great value to interdependence theory and to humans freely able to self-organize and make decisions; viz., like in our case study below on nuclear wastes.

3 Prior Research for Interdependence Theory

Our findings indicated that redundancy and emotion impede interdependence; that intelligence requires interdependence; that tradeoffs reflect uncertainty in a team's decisions; and that the determination of social context operates like a harmonic oscillator.

Redundancy contributes destructive interference (interdependence) to the operations of a team. It acts like an impurity in a crystal or in a team's structure, whereas the perfect team acts like a pure crystal (e.g., the orthogonal members of a team performing a military mission, a research project, or the conflict roles in a courtroom).

Consider Von Neumann's (p. 76, in [15]) attempt to construct a theory of self-replicating automata with thermodynamics from an individual automata's perspective. He concluded that it was not possible to choose the parts of a self-replicating automata in the right order. In contrast, we have argued that a team's perfect fit occurs when it minimizes the team's structural entropy sufficiently to allow a team to achieve maximum entropy production (MEP), demarcating good from bad teams.

When the perfect teams is constructed, information about how it functions as a team is lost; in agreement with what our model has found [3], Schrodinger [2] suggests that the perfect team loses information from its reduced degrees of freedom:

> … those true lovers who, as they look into each other's eyes, become aware that their thought and their joy are numerically one, not merely similar or identical …

Oppositely, when a team goes through a divorce, it requires energy to rip apart its structure (e.g., preceding the ViacomCBS merger when the two firms openly shared hostilities; during this period, both companies lost money; in [16]).

Excited versus ground states. Along the paths of entropy expenditures, we separate a team conceptually into structure and mission. Structurally, we have postulated that a perfect team operates at its lowest entropy, its ground state (like a biological enzyme, in [17]), allowing the perfect team to direct most of its available energy flow to achieve MEP as it functions. Thus, the best teams operate at ground states, the worst at excited states (e.g., marital divorce; the civil war in Syria; hostile mergers).

Intelligence is needed by a team or society to self-organize sufficiently to overcome barriers, as with the superior patent productivity we found for well-educated researchers in Israel compared to poorly educated researchers in neighboring MENA countries[1] [3]. By comparison, in a non-competitive society where individuals are dependent upon others, an education is less important than the street-smarts necessary to survive, e.g., Somoa today typifies the uneducated "Backward March of Civilization" [18].

Tradeoffs: Preferences and self-interest conflicts imply the existence of an uncertainty principle. Tradeoffs underscore the need for intelligence to navigate barriers, e.g. the EU-UK negotiations for a new treaty imply "… difficult trade-offs because both sides say they want close economic ties but have conflicting agendas" (in [19]).

Social harmonic oscillators (SHOs). In our case study [3], the Department of Energy (DOE) had been granted authority in 2005 to renew the closure of two high-level radioactive waste tanks (HLW). As part of the law that allowed DOE at its Savannah River Site (SRS), SC, to close these and its remaining 47 HLW tanks, the Nuclear Regulatory Commission (NRC) was given oversight of HLW tank closure. The SRS Citizen Advisory Board (CAB) and State of SC supported DOE's decision to renew closure. But, each month DOE would propose a plan to close its two HLW tanks, NRC would make an objection, and DOE would revise and resubmit its plan, an oscillation that continued until the fall of 2011 when SC complained in public before the SRS-CAB that DOE was going to miss its legally mandated milestone. The SRS-CAB then demanded immediate closure; both DOE and NRC agreed, and the tanks were closed. Since then, unlike the consensus-seeking citizens at DOE's Hanford site where no HLW tanks have yet to close, tanks at SRS have been closed regularly.

DOE's determination to close the tanks was insufficient to set the context. The conflict between DOE, NRC and SC was also insufficient. It took the citizens to demand that DOE and NRC close the tanks, an establishment of context that continues.

[1] Middle Eastern North African countries.

A summary of what can be measured for interdependence theory: *N, dof,* comparative MEP and WEP between teams, structural entropy production [3]; imagination ([20]; e.g., implicit racism); orthogonal roles, beliefs, stories [21]; rotation from oscillations [3]; and the competition to determine social reality and uncover deception [1].

4 Similarities with Quantum Mechanics

Interdependence is constructive and destructive interference. It functions like entanglement. Redundancy interferes with a team's operations. Boundary maintenance of a biological molecule, organism or team is needed to reduce destructive interference and as a barrier to redundant team members [3].

Information loss from joining two interdependent (social) objects together, i.e., subadditivity, is similar to no-cloning (e.g., p. 77, in [22]), precluding the replication of perfect teams, redundancy, however, sheds light on the collapse of interdependence.

Excited versus ground states. Tradeoffs that capture uncertainty in decision-making, and introduce intelligence for social decision-making, part of a social harmonic oscillator, followed by the resistance to a judgement determines social reality [3].

5 Conclusion

Best guess model: Our model of reality allows predictions based on what reduces interdependence (e.g., redundancy).

To Schrödinger's [2] question, *What is life?*, he answers what "an organism feeds upon is negative entropy … [where] metabolism … [gives] a measure of order … the organism succeeds in freeing itself from all the entropy it cannot help producing while alive … [allowing life to exist at a] fairly low entropy level …"

Generalizing Arrow [5], in the absence of competition, or a scientist establishing a game theory's payoffs, or a military Commander Officer's intent, or, for example, the U.S. Federal Reserve changing the interest rate, etc., it is impossible for an autonomous individual or team to determine social reality [3]; e.g., in the Carter Page matter, the FISA[2] court's chief judge, Rosemary Collyer, has officially questioned the accuracy of the FBI's past surveillance requests, indicating that for every request to surveil by the Department of Justice, something is needed like a standing advocate who can challenge requests on behalf of the subjects the FBI wants to surveil to give FISA judges a counter-position to consider [23].

To Schrödinger's question whether "two souls are one," we postulate that the collectivism in a socialist country, like Cuba, blocks self-organization to enslave and force its citizens to generate the WEP that benefits their masters, versus a country of free individuals able to self-organize into teams to compete and foster the team decision-making that focuses entropy to produce the MEP that improves collective well-being.

Is it merely romantic, popular or religious to accept that the cooperation model of Axelrod (1984) or the rational collective choice model [14] increases the value of

[2] Foreign Intelligence Surveillance Court is a U.S. federal court established and authorized under the Foreign Intelligence Surveillance Act of 1978 (FISA).

meaning, but not survival or evolutionary outcomes? Is meaning even necessary, when the meaning of the successful quantum theory has eluded researchers after a century of fierce debate [24]? Natural selection, and the evidence we have assembled, appears not to care. Possibly, a single meaning of life cannot exist, evidenced by our proposed function of opposing world views, adding to Schrodinger's [2] question, *What is life?*

References

1. Lawless, W.F., Mittu, R., Sofge, D.A., Hiatt, L.: Introduction to the special issue, artificial intelligence (AI), autonomy and human-machine teams: interdependence, context and explainable AI. AI Mag. **40**(3), 5–13 (2019)
2. Schrodinger, E.: What is life? The physical aspect of the living cell. based on lectures delivered under the auspices of the Dublin institute for advanced studies at Trinity college, Dublin (1944). Accessed 19 Dec 2019
3. Lawless, W.F.: The interdependence of autonomous human-machine teams: the entropy of teams, but not individuals, advances science. Entropy **21**(12), 1195 (2019). https://doi.org/10.3390/e21121195(2019)
4. Tetlock, P.E., Gardner, D.: Superforecasting: The Art and Science of Prediction. Crown, New York (2015)
5. Arrow, K.J.: Methodological individualism and social knowledge. Am. Econ. Rev. **84**(2), 1–9 (1994)
6. Axelrod, R.: The Evolution Of Cooperation. Basic, New York (1984)
7. Friedman, G.: The Pressure on China, Geoploitical Futures. https://geopoliticalfutures.com/the-pressure-on-china/. Accessed 7 Dec 2019, 19 Nov 2019
8. Rovelli's, C.: Seven Brief Lessons on Physics reviewed by Garner, D., (22 March 2016), Book Review: Seven Brief Lessons On Physics Is Long On Knowledge. New York Times, New York (2016)
9. Zell, E., Krizan, Z.: Do people have insight into their abilities? A Metasynthesis, Perspect. Psychol. Sci. **9**(2), 111–125 (2014)
10. Kelley, H.H.: Personal Relationships: Their Structure and Processes. Lawrence Earlbaum, Hillsdale (1979)
11. Amadae, S.M.: Rational choice theory. Political Science And Economics, Encyclopaedia Britannica (2016). https://www.britannica.com/topic/rational-choice-theory
12. Kahneman, D.: Thinking, Fast and Slow. MacMillan (Farrar, Straus & Giroux), New York (2011)
13. Israel, J.I.: The Enlightenment that Failed. Ideas, Revolution, and Democratic Defeat, 1748–1830. Oxford University Press, Oxford (2020)
14. Mann, R.P.: Collective decision making by rational individuals. PNAS **115**(44), E10387–E10396 (2018). https://doi.org/10.1073/pnas.1811964115
15. Von Neumann, J.: Theory of self-reproducing automata. Burks, A.W. (ed.). IEEE Trans. Neural Networks **5**(1), 3–14 (1966). University Illinois Press
16. Mullin, B., Flint, J.: Viacom-CBS Deal Drama Was Worthy of the Fall Lineup. A disgraced former CeO. A newly minted mogul. Lines from The Godfather. The path to a media reunion was rocky, Wall Street Journal, (13 Aug 2019)
17. Udgaonkar, J.B.: Entropy in Biology, Resonance (2001). https://www.ncbs.res.in/sitefiles/jayant_reson.pdf. Accessed 23 Jan 2020
18. EB, Editorial Board (EB): The Backward March of Civilization. A measles outbreak in Samoa kills 60 due to lack of vaccinations, Wall Street Journal, (6 Dec 2019). https://www.wsj.com/articles/the-backward-march-of-civilization-11575676876. Accessed 7 Dec 2019

19. Norman, L., Fidler, S.: After Brexit, Fractured EU Faces New Challenges. Britain's departure from the EU has unified the other members of the bloc, but life beyond Brexit promises to expose divisions among them, Wall Street Journal, (14 Dec 2019)
20. Harari, Y.N.: Sapiens: A Brief History of Humankind. HarperCollins, New York (2015)
21. Shiller, R.J.: Narrative Economics: How Stories Go Viral and Drive Major Economic Events. Princeton University Press, Princeton (2019)
22. Wooters, W.K., Zurek, W.H.: The no-cloning theorem. Phys. Today **62**(2), 76–77 (2009)
23. Lucas, R., Scathing Report Puts Secret FISA Court Into The Spotlight. Will Congress, NPR, (22 Dec 2019). https://www.npr.org/2019/12/22/790281142/scathing-report-puts-secret-fisa-court-into-the-spotlight-will-congress-act. Accessed 22 Dec 2019
24. Weinberg, S.: Steven Weinberg and the puzzle of quantum mechanics, replies by Mermin, N.D., Bernstein, J., Nauenberg, M., Bricmont, J., Goldstein, S. et al. In response to: The Trouble with Quantum Mechanics from the, 19 January 2017 issue. New York Review of Books, (2017b, 4/6)

Verifying Automation Trust in Automated Driving System from Potential Users' Perspective

Jue Li(✉), Chun Meng, and Long Liu

College of Design and Innovation, Tongji University, Fuxin Road 281, Shanghai 200092, China
{lijue,1931959,liulong}@tongji.edu.cn

Abstract. Trust was recognized as a key element in automation as it relates to the system safety and performance. However, automation trust established in research of automation industry may not be used directly into automated driving system. In order to identify automation trust in automated driving, a semi-structed interview was conducted based on a systematic review of automation trust with potential users. Results show that factors related to the vehicle itself, specifically its reliability, had the greatest association with trust. And it is difficult to apply automation trust theory directly to human-automated vehicle interaction as the driving task itself is dangerous and user of automated vehicle (AV) are usually not skilled operators. This study provides a new lens for conceptualizing automation trust, which can be applied to help guide future research and design procedures that enhance driver–automation cooperation.

Keywords: Trust · Automation · Automated driving · Human-Vehicle interaction

1 Introduction

It is estimated that 90–95% of the traffic accidents can be attributed to driver violations and improper human actions [1]. In order to reduce the human error in driving behavior, various driving assistance systems are constantly being applied to cars, and the continuous improvement of the automation level has become a trend in the development of modern vehicles [2]. According to the deployment of AVs in recent years, replacing human in driving task by an automated system can reduce congestion, reduce driver's fatigue, increase road safety and improve fuel efficiency [3].

However, many users may harbor distrust based on their preconceptions. It may subsequently lead to disuse part or all the function of the AV. In this case, the driver may switch from automated driving to manual driving, which would lose the superiority of the automation. In the case where the driver still chooses to use the automated driving system despite his distrust, he may continue to monitor and intervene in the automated system with a high degree of concentration. It will result in the driver undertaking too much work that the system should undertake, which may cause the driver's situation

M. Zallio (Ed.): AHFE 2020, AISC 1210, pp. 31–38, 2021.
https://doi.org/10.1007/978-3-030-51758-8_5

awareness losing, even fail to respond to the danger in time. Therefore, misuse and disuse not only result in waste of technology, but also more likely to bring excessive "supervision load" and "cognitive prejudice" to the automated driving system for the driver, leading to bad man-machine collaboration and unexpected driving accidents. On the contrary, excessive trust in AV could lead to excessive reliance and further to misuse of it. In May 2016, in the first fatal car accident of AV, Tesla's automated system did not promptly provide an warning when the car was in danger, and the driver inside the car was an avid fan of self-driving vehicles of Tesla, relying entirely on the car he drove and eventually caused the dramatic accident. Victoria's research conducted by collecting videos of AVs showed that when following the new driving duties (supervision), the driver did not receive enough support from the system but showed complacency and excessive trust [4].

Much of the existing research on factors that guide human-automation interaction is centered around trust, a variable that influences the willingness of users to rely on automation in situations characterized by uncertainty [5], just as it does in interpersonal relationships. This one-way trust in automation is called automation trust in previous studies. Research on automation trust has receiving more and more attention as it was recognized as a key issue that could hinder the success of AVs and a key determinant of human use and dependence on automated systems [6].

2 Theory of Automation Trust

2.1 Development of Automation Trust

In the context of human-automation interaction, Parasuraman is one of the earliest researchers who proposed trust in automation, pointing out that trust, mental workload, and risk can influence automation use [7]. Wicks believe that improper reliance related to misuse and disuse depends on how well trust matches the true capabilities of automation. Supporting proper trust is critical to avoiding misuse and disuse of automation, just as in promoting effective relationships [8]. Lee and Moray proposed that the calibration, resolution, and specificity of trust describe mismatches between trust and automation capabilities. Calibration refers to the correspondence between people's trust in automation and the automation's capabilities [9].

It is worth noting that Lee and See provide an integrated review of early research to elucidate the role of trust in automation and proposed a method for defining appropriate trust in automation which emphasized that overtrust is poor calibration in which trust exceeds system capabilities; with distrust, trust falls short of the automation's capabilities, as shown in Fig. 1 [10].

These existing researches triggered numerous studies on specific factors related to trust in automation that have greatly expanded knowledge regarding the variability of trust. For example, updated factors affecting trust, such as age, gender, personality characteristics, moods, and automation design characteristics [11–13]. However, these studies either focus on the macroscopic human-automation interaction or focus on specific industrial systems, which are not targeted at driving systems.

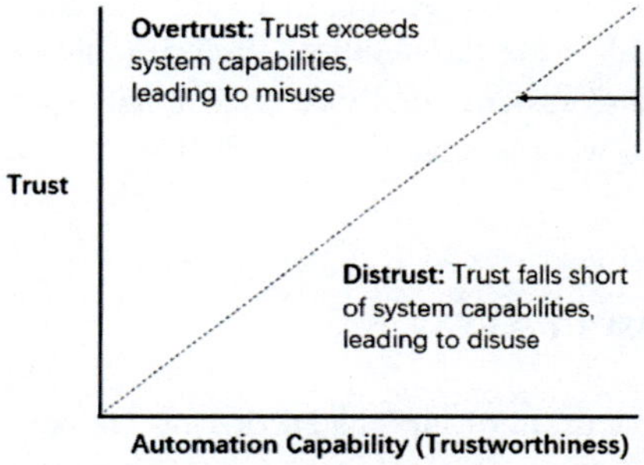

Fig. 1. The relationship among calibration, resolution, and automation capability in defining appropriate trust in automation by Lee and See.

2.2 Factors of Automation Trust

Identifying factors that affect automation and automation trust concept model is significant for calibrating trust and alleviating misuse of automation. There are no definite answers although there have been many studies focus on automation trust and have utilized a variety of different automation in diverse experimental paradigms to identify factors that impact trust. Therefore, we conducted a systematic review to identify those factors and models which can be used as verification material later.

This study retrieved academic papers in Web of Science core with the keywords "automation" and "trust", and 1060 academic papers were obtained. These papers were adopted for a co-citation analysis by the scientific metrics analysis tool *Citespace*, to obtain the highest co-cited papers.Of the 15 highest cited papers, 4 were of little relevance to automated trust and were excluded, and the rest were selected for a systematic literature review to identify the model of automation trust and the factors affecting automation

Table 1. Ten important factors of automated trust and their classification.

Category	Factors affecting automated trust
Human	Competency [5, 16, 19]
	Trust propensity [5, 15, 16]
	Personality Traits [5]
	Demographics [5, 14–17]
	Attitudes [5, 14, 15]
	Prior experience [5, 15, 17]
Vehicle	Reliability [1, 5, 14, 15, 18–20]
	Feedback [5, 14, 15, 19]
	information display [3, 7, 18, 19]
Environment	System environment [14–16]

trust. All the factors identified in the literature that affect the automation trust are summarized. The factors in Table 1 are the ten most cited. It indicated that the correlation between automation trust and system reliability is extremely high. And characteristics and experience of users, as well as environmental factors, can also affect automation trust.

3 Interviews on Trust in AVs

An interview with potential users of automated driving was conducted in Shanghai, China in November 2019, in order to understand the effectiveness of the automation trust theory in the field of driving. The definition of potential users in this paper is that they have a certain understanding of AVs but have not purchased or driven cars with automation functions for the time being. A total of 5 potential users were recruited for the interview, including 3 males and 2 females, aged between 24 and 35, all holding motor vehicle driving licenses.

The interview is divided into three parts, as shown in Table 2. The first part required participants to describe exactly what they know about driving and automated driving; the second part required participants to express their views and attitudes on AVs that have been deployed; the last part required participants to describe their trust in automated driving, and they were also asked to select the most important factors from the eight automation trust factors and comment on automation trust model and theories. All factors

Table 2. Interview Outline.

Interview stage	Interview focus
Part one	Driving experience with manual and automated driving
	Understanding of automated driving classification
	Understanding of AV functions
	Understanding of AV brands and models
Part two	Expectations of AV that have been deployed
	Concerns of AV that have been deployed
	Describe whether support for deploying AVs on Chinese roads
	Possibility of buying an AV
Part three	Describe their trust in automated driving
	Describe the most trusted form of automated driving (public transportation/private/logistics, etc.)
	Major causes of distrust/trust
	Select the most important ones from the given trust factors
	Comment on the automation trust model in context of automated driving

and conceptual model pictures that need to be selected or commented by the participants are printed on A4 paper and presented to the participants. The length of the interview is about 35 min.

4 Results and Discussion

4.1 Attitude Towards Automated Driving

Among the participants, four have a small amount of manual driving experience, and one has rich manual driving experience more than 5 years. Three of them clearly know the classification of AVs, one only knows that there is a classification standard for AVs, and one was only aware some news and brands of AVs.

About attitudes towards automated driving, the five participants affirmed the great value of automated driving in future road traffic, and were optimistic about the deployment of AVs in the next few years. Regarding what they concerned most about AV, one emphasized the importance of the safety of vehicles, also three emphasized that the new way of interaction between AVs and driver should be easy to learn. They also mentioned deployment scenarios, road infrastructure, driver training, and rights and responsibilities of accidents. When asking whether to buy an AV, although the participants all stated that they would not consider buying one in the next few years, most of them are very much looking forward to the opportunity to test drive or hire one. It indicates that most potential users may want to experience AVs at a low price and extreme security, rather than buying directly.

4.2 Trust in AVs

In general, participants in the interview trust in AVs to some extent, especially when imagining special-purpose scenarios in future road traffic, but there are still some hesitations in daily travel scenarios. Some of them mentioned that China's road traffic is complicated, so it is difficult to achieve large-scale applications of highly automated driving in the short term.

In terms of the correlation between driving experience and trust in AVs, the four participants who did not have much driving experience, especially two of them who showed higher concern about the development of automated driving, had greater trust in AVs than the one with five years of manual driving experience, which indicated that participants who know more about AVs and have less driving experience hold higher trust in AV. In addition, both users who actively mentioned the concept of trust in the second part of the interview have high trust in AVs. About whether they support deployment of AVs in China and how about the supported application scenarios, five participants in the interview expressed their support and expectation. Three participants are looking forward to the deployment of AVs in daily commuting, while the remaining participants showed relatively average support and expectations for different application scenarios. When talking about why they trust or distrust AVs, a female participant pointed out that the capability of AVs is difficult to define and presented to users accurately. She also believed that trust level in AVs should be reserved for the actual capability of the vehicle because of the high risk of driving tasks. Other participants agreed that the biggest reason for distrust of AVs is safety. One of them emphasized that the matching of road facilities with AVs was also his concern.

Participants were also asked to consider and comment on automation trust factors and conceptual models in the context of automated driving. Three participants with relatively low trust in automated driving believe that the most important factor affecting trust is the safety and reliability of the vehicle, which is consistent with the conclusions in the literature review of automation trust. And these two further believed that other factors are not worth mentioning, while the rest of the participants pay attention to all the factors. In addition, all five participants stated that it was difficult to make correct and valuable comments on those conceptual models of trust in the context of human-vehicle interactions because they have not actually driven AVs before. One of the participants who is very concerned about human-computer interaction said that she expects automated driving technology to solve some problems in manual driving, such as vehicle scratches caused by poor driving skills. She believes that users of AV who are not similar to pilots or operators of machine should not need much training, so several of the trust-related conceptual models provided in the interview could not accurately describe human interaction with AVs.

5 Conclusion

In order to clarify the similarities and differences between user trust in AVs and automated trust, theories of automation trust are discussed and compared with the trust in automated driving from the perspective of user comments in this study. The discussion and comparison of trust in automation and trust in automated driving in this study showed that trust is still one example of the important influence of affect and emotions on human-vehicle interaction. However, the human interacting with AV is usually an ordinary user with less training, rather than a skilled operator. It is difficult to ensure that every user knows and understands the capability of the system. In addition, the automated driving system requires users to hand over the driving right to the vehicle, which is a kind of entrust for life safety. Because of the danger of the driving task itself, users should trust the driving system more carefully.

This study provides a new angle of view for clarifying the variability of trust in automation. Its conclusions have implications for future research on automation trust, and have positive effects on developing training interventions that encourage appropriate trust in automated driving. However, due to the single research method and the small number of participants, it is difficult to draw universal conclusions to all the stakeholders of AVs. Furthermore, the current automation trust is derived almost exclusively via subjective response measured by a specific interaction with automation, and so does this study. However, in the interpersonal trust literature, physiological indicators and objective indicators such as the trust game of actual investment behavior are often used to measure trust [21]. And a person who reports trusting in automation may not show the same level of trust in action as there has always been a difference between an individual's self-report and his behavior. Therefore, future research should include empirical research on subjective and objective measurements in the context of human-automation interactions in order to expand the accuracy of current trust assessments and identify potential associations between these trust-related factors.

References

1. Rumar, K.: The role of perceptual and cognitive filters in observed behavior. Human Behavior and Traffic Safety. Springer, Berlin (1985)
2. Kyriakidis, M., De Winter, J.C.F., Stanton, N., Bellet, T., Van Arem, B., Brookhuis, K., et al.: A human factors perspective on automated driving. Theor. Issues Ergon. Sci. **20**(3), 223–249 (2019)
3. Verberne, F.M.F., Ham, J., Midden, C.J.H.: Trust in smart systems: sharing driving goals and giving information to increase trustworthiness and acceptability of smart systems in cars. Hum. Factors J. Hum. Factors Ergon. Soc. **54**(5), 799–810 (2012)
4. Banks, V.A., Eriksson, A., O"Donoghue, J., Stanton, N.A.: Is partially automated driving a bad idea? Observations from an on-road study. Appl. Ergon. **68**, 138–145 (2018)
5. Hoff, K.A., Bashir, M.: Trust in automation: integrating empirical evidence on factors that influence trust. Hum. Factors J. Hum. Factors Ergon. Soc. **57**(3), 407–434 (2015)
6. Mirchi, T., Vu, K.P., Miles, J., Sturre, L., Curtis, S., Strybel, T.Z.: Air traffic controller trust in automation in NextGen. Procedia Manuf. **3**, 2482–2488 (2015)
7. Parasuraman, R., Riley, V.: Humans and automation: use, misuse, disuse, abuse. Hum. Factors J. Hum. Factors Ergon. Soc. **39**(2), 230–253 (1997)
8. Wicks, A.C., Berman, S.L., Jones, T.M.: The structure of optimal trust: moral and strategic. Acad. Manag. Rev. **24**, 99–116 (1999)
9. Lee, J.D., Moray, N.: Trust, self-confidence, and operators adaptation to automation. Int. J. Hum. Comput. Stud. **40**(1), 153–184 (1994)
10. Lee, J.D., See, K.A.: Trust in automation: designing for appropriate reliance. Hum. Factors J. Hum. Factors Ergon. Soc. **46**(1), 50–80 (2004)
11. Pak, R., Fink, N., Price, M., Bass, B., Sturre, L.: Decision support aids with anthropomorphic characteristics influence trust and performance in younger and older adults. Ergonomics **55**(9), 1059–1072 (2012)
12. Stokes, C.K., Lyons, J.B., Littlejohn, K., Natarian, J., Speranza, N.: Accounting for the human in cyberspace: effects of mood on trust in automation. In: Proceedings of The 2010 International Symposium on Collaborative Technologies and Systems (CTS 2010), IEEE
13. Szalma, J.L., Taylor, G.S.: Individual differences in response to automation the five factor model of personality. J. Exp. Psychol. Appl. **17**(2), 71–96 (2011)
14. Hancock, P.A., Billings, D.R., Schaefer, K.E., Chen, J.Y.C., De Visser, E.J., Parasuraman, R.: A meta-analysis of factors affecting trust in human-robot interaction. Hum Factors J. Hum Factors Ergon. Soc. **53**(5), 517–527 (2011)
15. Schaefer, K.E., Chen, J.Y.C., Szalma, J.L., Hancock, P.A.: A meta-analysis of factors influencing the development of trust in automation: implications for understanding autonomy in future systems. Hum. Factors J. Hum. Factors Ergon. Soc. **58**(3), 377–400 (2016)
16. Merritt, S.M., Ilgen, D.R.: Not all trust is created equal: dispositional and history-based trust in human-automation interactions. Hum. Factors J. Hum. Factors Ergon. Soc. **50**(2), 194–210 (2008)
17. Beggiato, M., Krems, J.F.: The evolution of mental model, trust and acceptance of adaptive cruise control in relation to initial information. Transp. Res. Part F. Traffic Psychol. Behav. **18**, 47–57 (2013)
18. Dzindolet, M.T., Peterson, S.A., Pomranky, R.A., Pierce, L.G., Beck, H.P.: The role of trust in automation reliance. Int. J. Hum Comput Stud. **58**(6), 697–718 (2003)

19. Wang, L., Jamieson, G.A., Hollands, J.G.: Trust and reliance on an automated combat identification system. Hum. Factors J. Hum. Factors Ergon. Soc. **51**(3), 281–291 (2009)
20. Ben-Ner, A., Halldorsson, F.: Trusting and trustworthiness: what are they, how to measure them, and what affects them. J. Econ. Psychol. **31**(1), 64–79 (2010)
21. Chen, J.Y.C., Terrence, P.I.: Effects of imperfect automation and individual differences on concurrent performance of military and robotics tasks in a simulated multitasking environment. Ergonomics **52**, 907–920 (2009)

Human Interaction with Manned and Unmanned Aerial Vehicles

Differences Between Manned and Unmanned Pilots Flying a UAV in the Terminal Area

Anna C. Trujillo(✉), Roy D. Roper, and Sagar Kc

MS 152, NASA Langley Research Center, Hampton, VA 23681, USA
{anna.c.trujillo,sagar.kc}@nasa.gov, royroper@hotmail.com

Abstract. Detect and avoid (DAA), an essential component to integrate unmanned aircraft (UA) systems into the National Airspace System, focuses largely on developing and enhancing algorithms to assess and define requirements for the loss of well clear with other aircraft in the system. This flight simulation experiment focuses on terminal area alerting capabilities in and around the local airport traffic pattern and seeks to address under which conditions should the DAA system switch between large and small alerting criteria. Piloting differences observed between manned-aircraft and UA pilots while operating a UA in the terminal area in flight simulation are reported in this paper. Data indicate that UA pilots were more comfortable with smaller separations between their UA and other aircraft in the traffic pattern than the manned-aircraft pilots.

Keywords: Unmanned aircraft systems · Terminal area · Aircraft separation

1 Introduction

Detect and avoid (DAA) is an essential component of integrating unmanned aircraft systems (UAS) into the National Airspace System (NAS) because separation assurance to ensure safety is a major characteristic of the national airspace. Current concept of operations for UAS requires that UAS fly IFR (instrument flight rules) plans [1]. However, UAS will still need to self-separate when air traffic controllers are not providing that service—such as at non-towered airports with instrument approaches. In these situations, the ability to "see and avoid" other aircraft translates to "detect and avoid" (DAA) capability for UAS by necessity.

As part of NASA's UAS Integration in the NAS project, this experiment focuses on terminal area alerting capabilities and alert volume definition in and around local, uncontrolled airports for the UAS operator [2, 3], to amplify the en route minimum operational performance standards (MOPS) [4, 5]. This experiment begins to define the MOPS for UAS to land at an uncontrolled airport.

The overall goal of this research is to "[r]efine the well-clear definition and alerting requirements … in towered and non-towered terminal airspace" [6]. In particular, this experiment seeks to address under which conditions the DAA system should switch between large and small alerting criteria. En route DAA well clear (DWC) and alerting

M. Zallio (Ed.): AHFE 2020, AISC 1210, pp. 41–48, 2021.
https://doi.org/10.1007/978-3-030-51758-8_6

times were developed in Refs. [4, 7, 8]. For the terminal area, the DWC parameters were developed and defined in Refs. [2, 3], and the alert time was developed and defined in Ref. [9]. As a design artifact of this simulation experiment, both manned-aircraft pilots and pilots flying primarily unmanned aircraft (UA) were tasked to fly a UA in the terminal area. This paper addresses the piloting differences observed between the two pilot types while performing the task. Results regarding the DWC size, DAA terminal area (DTA) geometry and switching method from this experiment are available in Ref. [10], and a detailed discussion regarding the fine tuning of the alert times in the terminal area is available in Ref. [11].

2 DAA Well-Clear Definition

In order to fully understand the observed piloting differences between manned-aircraft and UA pilots, a basic understanding of the DWC and loss of well clear (LoWC) are needed. LoWC is defined as

$$\left[0 \leq \tau_{\text{mod}} \leq \tau^*_{\text{mod}}\right] \text{ and } \left[HMD \leq HMD^*\right] \text{ and } \left[-h^* \leq d_h \leq h^*\right] \tag{1}$$

where τ_{mod} is the temporal separation, *HMD* is the horizontal miss distance, and d_h and h^* are the vertical separation. Their defined maximum values for DAA are detailed in Table 1 and Ref. [4].

Table 1. DWC parameter values by DWC environment.

DWC	Parameter	Value
En route	τ^*_{mod}	35 s
	HMD^*	6076 ft
	h^*	450 ft
Terminal	τ^*_{mod}	0 s
	HMD^*	1500 ft
	h^*	450 ft

The DWC serves as a protective bubble around the UA. From (1), aircraft are "well clear" as long as they are outside of the minimum thresholds τ^*_{mod}, HMD^*, and h^*. If a LoWC is projected, then an alert is presented to the UA pilot depending on the expected time to LoWC (Table 2).

3 Experiment Design

Subject pilots flew a desktop simulation acting as a remote pilot of a simulated Predator UAV using the Vigilant Spirit Control Station (VSCS) [12]. While on a standard RNAV

Table 2. Alerting times and expected LoWC by DWC.

DWC	Alert (sec)		
	Preventive	Corrective	Warning
En route	60	60	30
Terminal	[a]	[a]	30

[a] No preventive or corrective alerts for terminal DWC.

GPS (area navigation global positioning system) approach to an uncontrolled Class E airport, the operator experienced a traffic conflict due to traffic pattern movements while multiple aircraft were operating in the vicinity. Experiment trials considered two DTA shapes—cylinder and prism—and two switching methods—intruder- and ownship-centric. In addition to encounter data, subjective feedback was requested from each subject regarding comfort or unease with each tested geometry. The background vehicles flew on predefined paths ensuring repeatability between subjects and only one predefined intruder vehicle per trial caused an alert.

Over 2 days, 9 subjects saw 44 trials each—evenly split between days. First day trials repeated on the second day, such that each subject saw every trial twice. Trials were grouped into sets by DTA shape and switching method, which repeated in a different order on the second day. No subjects recognized the repeated trials.

For this data analysis, there were 18 encounters: 4 on downwind, 8 on turn to base, 1 on base entry, and 4 with transiting aircraft. Eleven encounters resulted in a LoWC and seven did not result in a LoWC. There were also 11 en route alerts, 4 terminal alerts, and 3 encounters that had no alerts.

Subjects were asked to refrain from maneuvering their UA until an alert occurred unless they felt unsafe; in which case, maneuver as they saw appropriate. This was requested to have subjects accurately rate their preferences on the DTA geometry and switching method.

3.1 Pilot Type Characteristics

Four manned-aircraft pilots and five UA pilots (Table 3) flew the UA using the VSCS. The manned-aircraft pilot characteristics were either corporate, regional, or instrument instructor pilots with glass cockpit experience who had flown within the past year. The UA pilots had UAS operator experience and an instrument rating who had flown both manned an UA aircraft within the past year. Note that the unmanned subjects were primarily from the same company and typically flew UAS in military airspace.

3.2 DTA Switching

The three DTA/switching possibilities were (1) cylinder ownship, (2) cylinder intruder, and (3) prism intruder. Note that prism DTA geometry with ownship switching was not tested since the UA would cross into the terminal DWC near short final when the intruder would have already landed.

Table 3. Subject pilot type information.

Primary pilot type	Manned		Unmanned	
	Hours	Years	Hours	Years
Unmanned[a]	1158 ± 1244[c]	11.6 ± 7.6	923 ± 1093	2.1 ± 1.2
Manned[b]	2348 ± 1561	12.6 ± 7.6		

[a] 5 subjects
[b] 4 subjects
[c] ± 1 standard deviation

The general shape of the cylinder DTA was a 4.4 nautical mile (nmi) radius cylinder (typical of Class D airspace) centered on an airport's reference point, the approximate geometric center of all usable runway surfaces. The size and shape captured non-standard traffic patterns caused when using geographic reference points to navigate traffic pattern legs and allowed for a variety of runway configurations. However, the larger volume could include some transiting traffic.

The shape of the rectangular prism (prism) DTA encompassed a traffic pattern typical of light piston aircraft and was bounded 1 nmi from the runway centerline and 0.5 nmi from the runway threshold.

Cylinder Ownship. As the UA entered the cylinder ownship DTA, all intruder and background traffic switched to the smaller terminal DWC alerting scheme. If the UA were outside the cylinder ownship DTA, the larger en route DWC alerting scheme was applied to all intruder and background traffic.

Cylinder Intruder. For this DTA, the terminal DWC alerting was only applied by the UA to vehicles inside the cylinder DTA. For aircraft outside the DTA, the en route DWC alerting scheme was applied by the UA.

Prism Intruder. For this DTA, intruder switching functioned the same as the cylinder intruder DTA. Also, this DTA's smaller size meant that higher performance aircraft on a wider traffic pattern would not cause an en route DWC alert under normal conditions.

3.3 Intruder Aircraft Type and Speed

Intruder aircraft types (intruder) were divided into two groupings of speeds and VFR traffic pattern downwind offsets, piston and turbine. Piston intruder aircraft had speeds of 60 or 110 KTAS (knots true airspeed) (piston/low and piston/high, respectively) and flew the downwind leg of the VFR traffic pattern with a centerline runway offset of 4000 ft. Turbine intruder aircraft flew either 110 or 150 KTAS (turbine/low and turbine/high, respectively) and flew the downwind leg at a centerline offset of 8000 ft.

4 Results

Data were analyzed using IBM® SPSS® Statistics[1] version 25. Statistical significance was set at $p \leq 0.05$ with independent variables of DTA geometry and switching (cylinder ownship, cylinder intruder, or prism intruder), intruder type and speed (piston/low, piston/high, turbine/low, or turbine/high), and pilot type (UAS or manned).

This paper addresses the differences observed between manned and UAS pilots. Additional analyses on overall general results are available in Ref. [11] and further analyses regarding additional hypotheses are still underway.

4.1 Separation at Closest Point of Approach

Both horizontal and vertical closest point of approach (CPA) distances between the UA and intruder were recorded in addition to any LoWC. At LoWC, guidance bands appear on the map display indicating direction of turn, ascent, or descent to minimize time in LoWC.

For vertical separation at CPA, UA pilots had a smaller separation than manned pilots (666 ± 29 ft (mean ± 1 standard error of the mean) and 842 ± 40 ft, respectively) ($F_{(1,311)} = 13.2$; $p \leq 0.01$).

After each encounter, subjects were asked "[h]ow appropriate was your distance to the intruder at time of alert?" on a 100-point scale where 50 indicated the alert occurred when the intruder was acceptably distanced away. UA pilots indicated that distance was acceptable (47 ± 2) but manned pilots indicated the intruder distance was too close at the time of the alert (38 ± 2) ($F_{(1,153)} = 5.308$; $p \leq 0.03$), suggesting manned pilots felt the vertical separation from the intruder vehicle was less adequate.

4.2 Alert Duration

The duration of en route warning alert was significant for pilot type by DTA/switching interaction ($F_{(1,79)} = 5.855$; $p \leq 0.02$). As seen in Table 4, UA pilots had essentially the same en route warning alert duration for both DTA/switching types, which indicates UA pilots do not delineate between different parts of the approach to land. For manned pilots, the en route warning alert duration was higher with the prism intruder than for the cylinder ownship DTA. This may be due to pilot decision-making processes, where the manned pilot delineated between that length of the instrument approach, which allows for action to perform a missed approach, and the commitment to land at the final approach fix. Due to the size of the cylinder ownship DTA, warning alerting occurred while the UA vehicle was further away from the airport; therefore, the subject was less committed to land and immediately acted on the missed approach to avoid conflict. In the case of the prism intruder DTA, the opposite is true. The manned pilot had mentally committed to land and the alerting occurred closer to the runway, which forced a decision-making conflict and led to greater decision-making times. Note, for the cylinder intruder DTA, warning alerts only occurred under terminal DWC alerting.

[1] The use of trademarks or names of manufacturers in this report is for accurate reporting and does not constitute an official endorsement, either expressed or implied, of such products or manufacturers by the National Aeronautics and Space Administration.

Table 4. En route warning alert duration by pilot type and DTA/switching.

DTA/Switching	Mean (ft)	
	Manned Pilot	UA Pilot
Cylinder ownship	8.8 ± 0.7[a]	11.0 ± 2.6
Cylinder intruder	[b]	
Prism intruder	15.5 ± 1.8	12.2 ± 1.3

[a] mean ± 1 standard error of the mean
[b] No en route warning alerts

For recovery band duration, generally en route bands remained active longer than terminal bands. As seen in Table 5, UA pilots generally had a shorter duration than manned pilots for the recovery bands for both en route and terminal DWC ($F_{(1,39)} = 5.475$; $p \leq 0.03$ and $F_{(1,11)} = 6.952$; $p \leq 0.03$, respectively).

Table 5. Horizontal separation at CPA by pilot type and A/C type and speed.

DTA	Mean (sec)	
	Manned Pilot	UA Pilot
En route	27.3 ± 4.1[a]	22.1 ± 3.0
Terminal	19.7 ± 2.8	11.1 ± 1.9

[a] mean ± 1 standard error of the mean

En route LoWC duration was significant for pilot type by DTA/switching and pilot type by A/C type and speed. In general, UA pilots had a longer LoWC for cylinder DTAs than manned pilots ($F_{(2,120)} = 4.592$; $p \leq 0.02$) with the prism intruder LoWC duration not being dependent on pilot type (Table 6). Furthermore, UA pilots had a longer LoWC duration for turbine aircraft than manned pilots ($F_{(2,120)} = 4.120$; $p \leq 0.02$), which indicated manned pilots exhibited greater sensitivity to vehicle distinctions than the UA operator (Table 6).

4.3 Maneuvering

The recommended maneuver for the UA to avoid a LoWC while on IFR final approach was to initiate a missed approach. The missed approach command time (from trial initiation) was significant for pilot type ($F_{(1,186)} = 5.482$; $p \leq 0.02$). UA pilots had a significantly higher missed approach command initiation time (142 ± 4.9 s) than manned aircraft pilots (126 ± 4.9 s). In addition to initiating the missed approach earlier, manned pilots also felt pressure to maneuver before any bands appeared (34 ± 3) whereas UAS pilots felt that the bands appeared at about the time they wanted to maneuver (53 ± 2) ($F_{(1,124)} = 6.396$; $p \pm 0.02$) where 50 indicates pressure to maneuver was felt when the bands appeared. Lastly, when there was no LoWC, 76% of UA pilots felt no pressure to maneuver versus 24% of manned pilots ($X^2(1, N = 91) = 24.28$; $p \leq 0.01$).

Table 6. En route LoWC duration by pilot type and DTA/switching and aircraft type and speed.

DTA/Switching	Mean (sec)	
	Manned Pilot	UA Pilot
Cylinder Ownship	24.1 ± 2.2[a]	27.7 ± 2.7
Cylinder Intruder	21.7 ± 2.9	27.0 ± 1.6
Prism Intruder	19.8 ± 3.4	19.6 ± 2.5
A/C Type and Speed		
Piston/Low	27.4 ± 6.8	27.2 ± 3.5
Piston/High	12.5 ± 1.5	[b]
Turbine/Low	19.9 ± 2.2	24.2 ± 2.0
Turbine/High	22.4 ± 2.6	25.4 ± 1.3

[a] mean $\pm$ 1 standard error of the mean
[b] no data

5 Conclusions

In general, UA pilots were more comfortable in close proximity to other aircraft than manned-aircraft pilots. Manned pilots who were used to performing an instrument approach with a transition to out-the-window views at the final approach tended to delineate the approach into separate phases that led to differences in decision-making. For manned pilots, the decision to perform a missed approach occurred more quickly when farther away from the landing zone (where pilots were primed to discontinue the approach) but more slowly closer to the runway (where the manned pilot expected to land). UA pilots exhibited greater uniformity in decision-making, being used to a ground control station monitor that remained a constant throughout the approach and landing phase.

All vehicles except the UA within the simulation used the same aircraft icon, for which the UA pilots tended to treat most vehicles as the same, typically only taking note of vehicle speeds. Manned pilots, though, made clear distinctions between piston and turbine vehicles types, speeds, and associated pattern distances which resulted in more immediate action and lower LoWC durations for conflicting turbine aircraft than UA pilots. Interestingly, manned pilots tended to maintain a typical average separation distance for all aircraft types and speeds with the exception of the low speed turbine, while UA pilots exhibited non-uniform distances among all conflicting aircraft types and speeds. Manned pilots also initiated missed approaches sooner than UA pilots, which may indicate manned pilot discomfort to intruder distances, or UA pilots taking more time to observe vehicle trends, or both.

These results indicate that there are piloting differences between UA and manned-aircraft pilots with the primary differences being that UA pilots are comfortable being closer to other aircraft in the terminal area. What fully drives this comfort level for closer proximity has not been fully studied. Further research should be conducted to determine how these differences would affect the safe integration of UAS into the NAS.

Acknowledgments. This research was funded by the Detect and Avoid Well-Clear subproject of NASA's UAS Integration in the NAS project. Tod Lewis and Michael Vincent contributed to the design of the experiment. For simulation, the authors would like to thank Dimitrios Tsakpinis, simulation development lead, Robert Myer, Joel Ilbuoudo and Joshua Kaurich (SAIC), and Kristen Mark and Anna DeHaven (Craig Technologies). During data collection, the authors would like to thank Paul Volk of AAG, and NASA Langley's Air Traffic Operations Lab—Ed N. Scearce, NASA, Chad Chapman, David West of SSAI, and Troy Landers and Joe Mason of Metis Technology Solutions.

References

1. RTCA, Inc.: UAS Landing Requirements. Requirements from SC-228 committee meeting. SC-228 (2018)
2. Trujillo, A.C., Jack, D.P., Tsakpinis, D.: En route detect and avoid well clear in terminal area landing pattern (AIAA 2018–2872). In: 2018 Aviation Technology, Integration, and Operations Conference, AIAA Aviation Forum American institute of aeronautics and astronautics, Atlanta, GA, p. 11 (2018)
3. Vincent, M.J., Trujillo, A.C., Jack, D.P., Hoffler, K.D., Tsakpinis, D.: A recommended DAA well-clear definition for the terminal environment (AIAA 2018–2873).In: 2018 Aviation Technology, Integration, and Operations Conference, AIAA Aviation Forum American Institute of Aeronautics and Astronautics, Atlanta, GA, p. 13 (2018)
4. RTCA Special Committee 228: DO-365: Minimum Operational Performance Standards (MOPS) for Detect and Avoid (DAA) Systems. Standards, RTCA, Inc., (2017)
5. RTCA Special Committee 228: DO-366: Minimum Operational Performance Standards (MOPS) for Air-to-Air Radar Detect and Avoid (DAA) Systems Phase 1. Standards, RTCA, Inc., (2017)
6. Vincent, M.J., Roper, R.D.: Concept of operations for UAS detect-and-avoid in terminal operations. In: NASA (ed.). NASA Langley Res. Cent. Hampton, VA (2018)
7. Ghatas, R.W., Jack, D.P., Tsakpinis, D., Vincent, M.J., Sturdy, J.L., Muñoz, C.A., Hoffler, K.D., Dutle, A.M., Myer, R.R., DeHaven, A.M., Lewis, E.T., Arthur, K.E.: Unmanned aircraft systems minimum operational performance standards end-to-end verification and validation (E2-V2) simulation. In: Administration, N.A.a.S. (ed.), NASA Langley Res. Cent. Hampton, VA, p. 124 (2017)
8. Muñoz, C.A., Narkawicz, A., Hagen, G., Upchurch, J., Dutle, A., Consiglio, M.C., Chamberlain, J.P.: DAIDALUS: detect and avoid alerting logic for unmanned systems. In: 2015 IEEE/AIAA 34th Digital Avionics Systems Conference (DASC), p. 18 (2015)
9. Rorie, C., Monk, K., Roberts, Z., Brandt, S.: Unmanned Aircraft Systems (UAS) integration in the National Airspace System (NAS) Project: Terminal Operations HITL 1B Primary Results. In: NASA (ed.). NASA Ames Research Center (2018)
10. Jack, D.P., Hoffler, K.D., Lewis, E.T.: Terminal Operations HITL 2 (TOPS2) SC-228 Outbrief. NASA, NASA Langley Res. Cent. p. 49 (2019)
11. Jack, D.P.: Alert timing assessment for unmanned aircraft system terminal area operations. 2020 AIAA SciTech Forum. AIAA, Orlando, FL (2020)
12. Cooper, M.: AFRL shares UAV software to further research. Wright-Patterson AFB, p. 1 (2017)

Obtaining Public Opinion About sUAS Activity in an Urban Environment

Caterina Grossi[1], Lynne Martin[2](✉), and Cynthia Wolter[1]

[1] San Jose State University, San Jose, CA, USA
Caterina.Grossi@sjsu.edu, Cynthia.Wolter@NASA.gov
[2] NASA Ames Research Center, Moffett Field, CA, USA
Lynne.Martin@NASA.gov

Abstract. Members of the public completed a short survey to complement a larger flight demonstration and data collection exercise in Corpus Christi, Texas. Respondents were invited to share their concerns and were asked about their knowledge of small Unmanned Aerial System (sUAS) operations. Participants were familiar with sUAS operations and reported feeling moderately comfortable with urban area operations, reflecting this in opinions that sUAS are as safe as other modes of transport. Compared to other UAS opinion surveys, participants gave similar responses to affect questions but were more knowledgeable concerning sUAS operational regulations. At first exposure to these live sUAS flights, participants reported positive impressions about the traffic management system that was being demonstrated.

Keywords: Unmanned aerial systems · UAS traffic management · Autonomy · Public opinion

1 Introduction

Interest in small unmanned aerial systems (sUAS) use by government, commercial, and recreational users has increased significantly over the past decade [1, 2]. Due to this increased interest, many studies have examined the general public's perception of sUAS use and its relation to concepts such as privacy, safety, and trust [3–7]. Through a thematic analysis of city council meetings from over 20 cities between 2014–2017, Nelson and Gorichanaz [4] found six major areas of public concern regarding sUAS: privacy, safety, enforceability, crime, nuisance, and professionality. Overall, these themes suggest that trust and transparency in sUAS are needed for the general public to accept the use of sUAS in urban environments. Additionally, a common factor found to correlate with the public's perception of sUAS and the themes derived from Nelson and Gorichanaz [4] is the public's knowledge of sUAS regulations and technology [8, 9].

To address the question of safety, and ultimately trust in sUAS technology, the Unmanned Aircraft Systems (UAS) Traffic Management (UTM) research project has developed and tested concept ideas for enabling sUAS operations in low altitude airspace

M. Zallio (Ed.): AHFE 2020, AISC 1210, pp. 49–55, 2021.
https://doi.org/10.1007/978-3-030-51758-8_7

(from the ground to 400 feet). A series of flight test demonstrations were organized over five years, with one of the final and most complex flight tests conducted in August 2019, in the city of Corpus Christi, Texas, USA. This testing resulted in over 400 data collection flights using eight live rotorcraft, with nine flight crews flying pre-planned scenarios in the urban downtown and waterfront areas of this city. Test scenarios were designed to include a variety of elements, including live and simulated vehicles with different mission objectives, and personnel in a variety of roles, including flight crews and mission support.

Because UTM facilitates transparency in sUAS operations, both between operators and to the public, the purpose of this paper was to examine the relationships between the general public's knowledge of sUAS, perceived sUAS risk, and their perception of UTM. To obtain a reference point that captured the public's perception of the flight test operations, a survey was developed and administered to members of the public present during the sUAS activities. It was predicted that general knowledge about sUAS operations and perceived risk of sUAS operations would be negatively correlated, with sUAS knowledge and perception of UTM showing a positive correlation. Additionally, it was predicted that the presence of a UTM system would lead to a decrease in public concern about sUAS and an increase in trust and perceived safety of sUAS operations.

2 Method

Participants. Forty-five people completed the online public opinion survey. Participants volunteered to take the survey, creating a self-selected sample. No demographic data was collected from participants, allowing responses to be as anonymous as possible. Thus, the gender and ages of the participants are unknown. The common characteristic shared by survey respondents is that they were all in the Corpus Christi downtown and waterfront area during a phase of the flight test. Seven respondents completed the survey following a Town Hall meeting run by the test site prior to any live flights, 27 completed the survey during the first week of testing and data collection, and a further 11 completed the survey during the second week of testing.

Approach. Visual Observers were stationed at points around the perimeter of the urban flight test area, primarily to report sUAS locations back to flight crews, but also to serve as public relations and interact with the public in the vicinity of the live operations. They invited members of the public who showed interest in the flight test to complete an online survey. Another role for these team members was to explain the goals of the flight tests and to answer general questions from the public.

Survey. The survey included a total of ten questions. Four questions were modeled after the UAS public opinion work completed by Clothier et al. [9] asking participants about their level of comfort and concerns with sUAS activity. Three questions were general knowledge questions, asking about current (as of summer 2019) restrictions on sUAS operations, similar to [8]. The final three questions asked participants' opinions on a UTM system, whether this would increase their confidence in urban sUAS activity, what information they would like to be able to obtain about said activity, along with the method they would prefer to gain that information. Four of the questions involved rating opinions on a seven-point scale from 1 as "low" or "less", to seven as "more or "a great deal". Four questions were single or multiple-choice, two were true/false, and one was a free-response question. On average, the survey took most participants less than five minutes to complete.

Data collection. Forty-five responses were recorded for the online survey. These responses were then analyzed, and participant comments were listed and compared.

3 Results

3.1 Descriptive Analysis

Public Concerns. The first section of the survey consisted of four questions aimed at identifying the main concerns of the general public related to sUAS. Respondents reported feeling moderately comfortable (M = 4.25, SD = 1.99) with urban sUAS operations. In contrast, respondents reported feeling mildly concerned (M = 3.29, SD = 1.82) about sUAS use in their community. A significant difference was not found between data collected prior to and during live operations for either of these items (F = 1.958, p = .192; F = .866, p = .450).

When asked about their top concern for sUAS technology, 43% of respondents (n = 19) reported it to be privacy, while 27% of respondents (n = 12) cited safety to those on the ground as a primary concern. Visual and noise disturbances were among the least cited concerns (n = 3, n = 1). Finally, 16% of respondents (n = 7) report having no concerns about UAS technology at all.

The final two questions in the "concern" section of the questionnaire centered around the public's view of sUAS safety when compared to (a) other aerial vehicles and (b) other cargo and goods transportation vehicles (i.e., delivery trucks). On average, respondents reported feeling that sUAS operations are about the same level of safety as air (M = 4.35, SD = 1.48) and delivery (M = 4.5, SD = 1.61) vehicles. No significant difference was found between data collected prior to and during live operations for either item (F = .200, p = .824; F = 1.596, p = .278).

Public Knowledge of UAS. The goal of the second section of the questionnaire was to determine the extent of general public knowledge regarding sUAS, in addition to where they currently seek out this information.

First, a total of five true/false questions were asked to determine the level of general sUAS knowledge among the public. When asked if most sUAS are capable of operating completely autonomously, 46% of respondents (n = 19) correctly selected "false", with 19% of respondents (n = 8) selecting "true" and 35% selecting "don't know/unsure" (n = 14). Respondents were also asked about when FAA approval is necessary. Approximately half of respondents correctly selected that FAA approval is necessary to legally operate sUAS at night (n = 23) and over people (n = 21). Additionally, 29% of respondents (n = 11) correctly selected that FAA approval is not required to legally operate UAS over buildings, leaving 32% of respondents (n = 14) selecting "true" and 39% (n = 16) selecting "don't know/unsure". Finally, when asked if FAA approval is required to legally operate sUAS beyond visual line of sight, 46% of respondents ($n = 19$) correctly selected "true", with 15% ($n = 6$) selecting "false," and 39% ($n = 16$) selecting "don't know/unsure."

Data regarding where respondents usually obtain information about sUAS was also collected. Half of respondents reported that they obtain this information through general websites ($n = 21$, Fig. 1). Other commonly used sources included government websites ($n = 14$), social media ($n = 12$), and broadcast news ($n = 12$). Interestingly, 28% of respondents ($n = 11$) reported never having searched out information about sUAS.

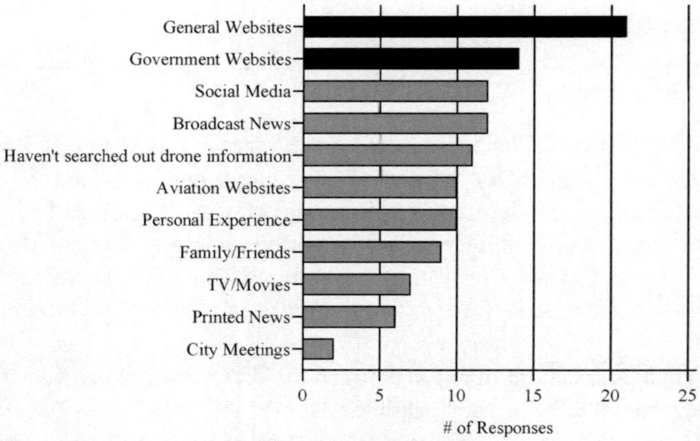

Fig. 1. Frequency of source used by participants for obtaining sUAS information.

Public Perception of UTM. The final section of the questionnaire focused on the public's perception of the UTM system and its impact on sUAS operations. When asked how a UAS traffic management system would impact the safety, trustworthiness, and concern of sUAS operations in urban areas, respondents reported, on average, that they thought UTM could make sUAS operations a little more safe ($M = 5.22$, $SD = 1.62$) and thought that UTM could help them have a little more trust in sUAS operations ($M = 5.14$, $SD = 1.50$) (Fig. 2). However, the increase in trust did not coincide with a decrease in the level of concern as respondents reported that they would be as concerned about these operations as for operations without UTM ($M = 4.69$, $SD = 1.55$). Responses given after having watched live operations had lower ratings than those that came only after attending the town hall meeting before data collection began. A significant difference was found for perceived trust in sUAS using UTM ($F = 7.462$, $p = .045$), where those who had attended the town hall meeting thought they would trust the sUAS operations under UTM more than those who had seen the UTM operations. There were no significant differences found between groups for perceived safety in sUAS using UTM ($F = 2.385$, $p = .208$) nor concern about sUAS operations ($F = .029$, $p = .971$). Half of the respondents suggested that websites (government or other organizations) would be the best places to access information about sUAS operations ($n = 19$). Social media and news media were other popular suggested sources for this information ($n = 11$). Finally, respondents were asked what sUAS information they would like to be able to access through UTM. Ninety-two percent of respondents ($n = 36$) reported the purpose of the sUAS flight as desired information and 85% ($n = 33$) felt that flight location would be beneficial information to make accessible to the public.

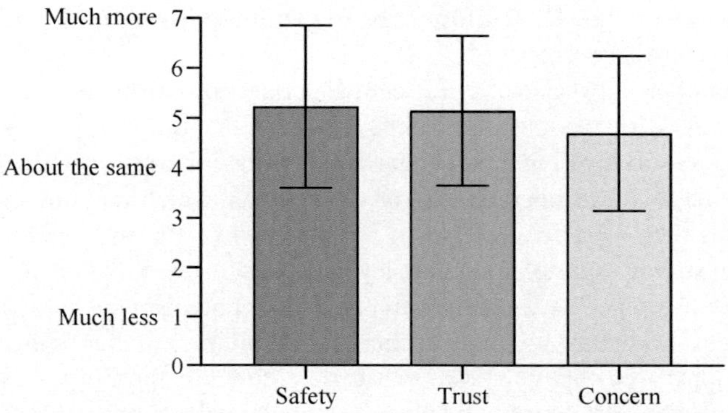

Fig. 2. Mean ratings of UTM's effects on perceived safety, trustworthiness, and concern regarding sUAS use in urban environments

Statistical Analysis. A Pearson's correlation was computed to examine the relationships between level of concern about sUAS operations, sUAS knowledge, and opinion of UTM. No significant correlation was found between sUAS concern and sUAS knowledge, $r(33)= .030$, $p = .866$, nor between sUAS knowledge and opinion of UTM, $r(29) = -.180$, $p = .333$. Additionally, although there was a positive relationship between sUAS concern and opinion of UTM, no significant correlation was found, $r(28) = .319$, $p = .085$.

4 Discussion and Conclusions

Previous studies regarding the public's opinion of UAS have focused on knowledge, privacy and risk concerns. As the purpose of this study was to compare public opinion during our broader flight-testing activities to general reported opinion, the public survey foci mirrored these themes.

Respondents to this survey displayed a reasonable level of knowledge of FAA regulations governing commercial UAS operations in the USA. For the five true/false questions, 45% of answers selected were correct, however, in many responses participants (55%) either admitted that they did not know the correct response (39%), or answered incorrectly (16%). When converted into a score, all responses combined had a mean of 1.64 out of 2. This compares with Keller et al.'s [8] results, where three true/false UAS knowledge questions were asked and they reported a mean value of 1.89 out of 2. Following the work reported by Nelson and Gorichanaz [4], the current participants were asked to indicate how they obtained their information about sUAS. Nineteen percent reported going online to look at general websites and a further 21% reported looking at other specific websites, giving a total of 40% of respondents reporting looking for information online. This is a slightly higher proportion of the sample compared to Nelson and Gorichanaz [4], who found 31% of their respondents looked for information online, although this may reflect a genuine increase in people turning to the web as a source of information. This proportion further increased when participants were asked where they would like to find sUAS information in the future, where 66% selected online media as their preferred source.

Keller et al. [8] also asked a set of four questions probing trust, reporting a relatively low level of trust from respondents. However, that question set asked respondents their willingness to travel in a passenger UAS, while the trust question in the current survey pertained to sUAS operations over urban areas, which may involve less severe consequences than a passenger-UAS if there were to be a mishap. The respondents in the current survey reported moderate levels of trust ($M = 5.14$) for sUAS urban operations, and the difference between the perceived risk of being below an sUAS versus occupying a larger UAS themselves may account for obtaining a higher rating than Keller et al. [8].

A third area of enquiry was to ascertain how safe the Corpus Christi public felt with urban sUAS operations taking place around them. Participants responded they felt sUAS safety was "about the same" as it was for other vehicles (both ground and air transport). Clothier, et al. [9] asked similar questions to explore the Australian public's assessment of UAS safety and also reports that respondents felt that UAS operations have similar safety to manned aircraft.

Herron et al. [10] and Nelson et al. [3] focused their studies on opinions of privacy with respect to UAS. They found that privacy was one of the greatest concerns for the public. The current results echo this, with privacy being the clear leader for concern within this small sample (with 43% of the respondents listing privacy as their primary concern). It is promising that participants reported their *level* of concern to be mild ($M = 3.29$), i.e., while privacy was a primary concern, respondents were not overly troubled about it. Those who attended the town hall meeting expressed more concern than those who saw the live operations, perhaps due to self-selection.

Previous studies have reported a correlation between participants' general knowledge about UAS and their acceptance of the technology, where lower levels of concern varied together with greater levels of trust (e.g., [4]). It was hypothesized that a similar relationship would apply to knowledge of sUAS and predicted opinions of the UTM system. Respondents reported, on average, that they thought UTM could make sUAS operations in urban areas a little more safe ($M = 5.2$) and that they perceived they would trust sUAS operations a little more ($M = 5.1$) if those operations were participating in a traffic management system. Comparing participants' knowledge, as gauged by their responses to the true/false fact questions, participant sUAS knowledge has a positive but non-significant relationship with ratings of trust in sUAS operations that are using UTM (r_s (35) = .323, p = .08). But there is no relationship between sUAS knowledge and perceptions of safety of sUAS operations that are using UTM (r_s (35) = .07, p = .88).

4.1 Conclusion

Although this survey study included only a small sample of people within a single region, responses on perceptions of risk, preferences for information and trust were similar to those found in earlier studies on public opinions of UAS operations by Clothier et al. [9], Nelson and Gorichanaz [4] and others. Furthermore, there was also a positive relationship found between participant knowledge and trust in the focus of sUAS traffic management technology. It was encouraging to find that after only minimal exposure to the UTM concept and after witnessing live flights utilizing the UTM system in their downtown, the public of Corpus Christi had a positive perception of these kinds of sUAS operations.

Acknowledgements. Many thanks to Natalia Menking (LSUASC, Corpus Christi, Texas) who spearheaded the data collection effort in the field and trained the Visual Observers.

References

1. Calo, R.: The case for a federal robotics commission. Brookings Institution (2014)
2. West, J., Bowman, J.: The domestic use of drones: an ethical analysis of surveillance issues. Public Adm. Rev. **76**(4), 649–659 (2016)
3. Nelson, J., Grubesic, A., Wallace, D., Chamberlain, A.: The view from above: a survey on the public's perception of unmanned aerial vehicles and privacy. J. Urban Technol. (2019). https://www.researchgate.net/profile/Jake_Nelson2/publication/330865366
4. Nelson, J., Gorichanaz, T.: Trust as an ethical value in emerging technology governance: the case of drone regulation. Technol. Soc. (2019). https://www.researchgate.net/profile/Jake_Nelson2/publication/332644353
5. PytlikZillig, L.M., Duncan, B., Elbaum, S., Detweile, C.: A drone by any other name: purposes, end-user trustworthiness, and framing, but not terminology, affect public support for drones. IEEE Technol. Soc. Mag. **37**(1), 80–91 (2018)
6. West, J.P., Klofstad, C.A., Uscinski, J.E., Connolly, J.M.: Citizen support for domestic drone use and regulation. Am. Polit. Res. **47**(1), 119–151 (2019)
7. Zwickle, A., Farber, H., Hamm, J.: Comparing public concern and support for drone regulation to the current legal framework. Behav. Sci. Law **37**, 109–124 (2018)
8. Keller, J., Adjekum, D., Alabi, B., Kozak, B.: Measuring public utilization perception potential of unmanned aircraft systems. Int. J. Aviat. Aeronaut. Aerosp. **5**(3), 9 (2018)
9. Clothier, R., Greer, D., Greer, D., Mehta, A.: Risk perception and the public acceptance of drones. Risk Anal. **35**(6), 1167–1183 (2015)
10. Herron, K., Jenkins Smith, H. and Silva, C.: US public perspectives on privacy, security, and unmanned aircraft systems. Cent. Risk Crisis Manage. Univ. of Oklahoma (2014)

Learn on the Fly

Yang Cai(✉)

Visual Intelligence Studio and Cylab Institute, College of Engineering and School of Computer Science, Carnegie Mellon University, Pittsburgh, PA 15213, USA
ycai@cmu.edu

Abstract. In this study, we explore the biologically-inspired Learn-On-The-Fly (LOTF) method that actively learns and discovers patterns with improvisation and sensory intelligence, including pheromone trails, structure from motion, sensory fusion, sensory inhibition, and spontaneous alternation. LOTF is related to classic online modeling and adaptive modeling methods. However, it aims to solve more comprehensive, ill-structured problems such as human activity recognition from a drone video in a disastrous environment. It helps to build explainable AI models that enable human-machine teaming with visual representation, visual reasoning, and machine vision. It is anticipated that LOTF would have an impact on Artificial Intelligence, video analytics for searching and tracking survivors' activities for humanitarian assistance and disaster relief (HADR), field augmented reality, and field robotic swarms.

Keywords: AI · Machine learning · Drone · UAV · Video analytics · SLAM · HADR

1 Introduction

We often do things "on the fly" in everyday life. We gain experience without preparation, responding to events as they happen [1]. We often learn new things in that way. For example, children learn to walk, talk, and ride a bike on the fly. Historical examples include Neil Armstrong landing the lunar module on the moon. Apollo 13 crews managed to return to Earth after an explosion. Network administrators responded to the first computer worm created by Robert Morris. More recently, epidemiologists have been fighting the COVID-19 coronavirus outbreak based on live data.

Learn-on-the-fly (LOTF) is a way of *active learning* by improvisation under pressure. It is not about learning how to fly, but rather how to learn quickly in challenging situations that may be mobile, remote, and disastrous, where other data-centric *passive learning* methods often fail. LOTF is related to classic "online modeling" or "adaptive modeling" methods such as Kalman Filter, Particle Filter, Recursive Time Sequence Models, and System Identification, which adapt to dynamic environments. LOTF aims to tackle more robust, complex problems such as human activity recognition from a drone video in a disastrous environment.

M. Zallio (Ed.): AHFE 2020, AISC 1210, pp. 56–63, 2021.
https://doi.org/10.1007/978-3-030-51758-8_8

In addition, LOTF aims to build explainable AI models that enable human-machine teaming (including visual representation and visual reasoning) toward machine vision-like humans. LOTF can also incorporate lightweight machine learning algorithms such as Bayesian network. In this paper, the author overviews biologically-inspired LOTF algorithms in non-technical terms, including pheromone trails, structure from motion, sensory fusion, sensory inhibition, and spontaneous alternation. It is anticipated that LOTF will have an impact on artificial intelligence, in particular, video analytics for searching and tracking survivors' activities for humanitarian assistance and disaster relief (HADR), augmented reality, and robotic swarms.

2 Pheromone Trails

It has long been known that social insects such as ants use pheromones to leave information on their trails for foraging food, leaving instructions for efficient routes, for searching, and for making recommendations. Similarly, Amazon's retail website suggests similar products based on the items in a user's online shopping cart. In practice, the term "pheromone" proves useful in describing behaviors such as trail formation in a sequence of spatial and temporal data.

The generalized pheromone update model can help us to discover motion patterns in videos, which transforms invisible patterns of moving objects into visible trails that accumulate or decay over time, much like a scent. Pheromones decay at a certain rate, thereby reducing the risk of followers repeating the same route. It also helps prevent them from reacting to a rapidly changing environment. Here, we generalize pheromone deposits and decay at the pixel level in two-dimension, where a "deposit" function is to add a unit of digital pheromone (in color) each time an object passes that pixel location until the value reaches its maximum. The "decay" function is to remove a unit of pheromone at a certain rate until the existing pheromone at the pixel location reaches zero. Figure 1 shows an example of traffic patterns over time from an aerial video. The heat map shows that the center lane has the heaviest traffic.

Fig. 1. The digital pheromones show the traffic flow over time from an aerial video

3 Structure from Motion

Motion perception is our instinct for survival. It is a vital channel for us to map our world. To extract the motion features, we can use Optical Flow [10] to describe motion, direction, and strength in terms of motion vectors. Optical Flow assumes the brightness distribution on moving objects in a sequence of images is consistent, which is referred to as "brightness constancy." We use the Horn-Schunck algorithm to minimize the global energy over the image. This algorithm generates a high-density of global optical flow vectors, which is useful for measurement purposes. We then use grid density to define the number of motion vectors in a frame. For example, we can plot a motion vector for every 10 pixels horizontally and vertically respectively.

Dynamic visualization of the field of optical flow is a critical component to reveal the changes of flow patterns over time. This is called a *flow map*. In addition to the flow map, we can visualize the motion vector in the color space of hue, saturation, and value (HSV), wherein hue represents the angle of the vector, and value represents the magnitude of length of the vector. The optical flow vector angle can be naturally mapped to hue in the HSV color system, both in range between 0 and 360 degrees. The magnitude of the vector can be mapped to a value between 0 and 1. Saturation value for this visualization is constant, so we can set it as 1 – the highest value by default. We chose the HSV color space for mapping the two parameters because of its simplicity. Figure 2 shows that the optical flow heat map visualizes the slow-moving utility truck in the wrong direction. This method is based on the assumption that the video is from a stationary camera. The heuristic algorithm for segmentation from a moving camera is in reference [9].

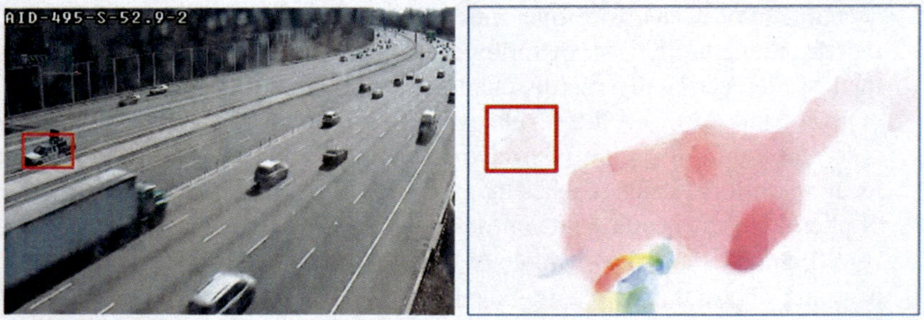

Fig. 2. The optical flow map visualizes the slow-moving utility truck in the wrong direction

Motion creates depth perception that can be used for reconstructing three-dimensional objects. Given a 2D video from a drone camera, we use Stereo-photogrammetry [11] to extract the 3D measurements. By analyzing the motion field between frames, the algorithm is designed to find corresponding points shared between frames, allowing for reconstruction of 3D structural coordinates from a single camera. The key assumptions of this method are the video contains enough high-contrast corner-like feature points, which are used for matching the corresponding structural features; and the geometric transformation caused by the motion is Homographic Transformation [12]. Note that stereo-photogrammetry is computationally intensive. We must down-scale the 4K video to a manageable size in order to achieve a reasonable computation time. Figure 3 shows the results of a 3D reconstructed archeological site in Paspardo in the Italian Alps.

Fig. 3. The archeological site is 3D reconstructed from a drone video with an RGB camera

For the last two decades, Structure-from-Motion (SfM) has been evolved into a popular technology for 3D imaging with an affordable single camera, a pair of stereo cameras, or multiple cameras [13]. The RGB camera-based SfM methods commonly need structural features such as Difference of Gaussian (DoG) SIFT features [14, 15] or FAST corner features [16] to match the structural features between frames in the video, and to calculate the homographic transformation matrix accordingly for Simultaneous Localization and Mapping (SLAM) [17]. Similar to Stereo-photogrammetry, the matching algorithm requires a minimal number of features in consecutive frames of the video. Unfortunately, in many cases, there are not enough matching features between frames, due to "featureless" smooth walls, blurry images, or rapid movement of the camera. Figure 4 shows results of the SLAM of the floor of an office building with a stereo camera, where the green dots represent the camera's motion path, and the other color dots represent the walls and the floor. The point cloud of the ceiling has been cut away to increase visibility.

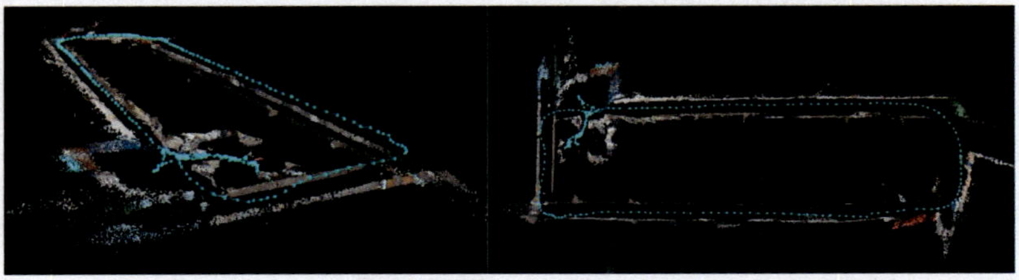

Fig. 4. The SLAM results of a floor plan and path in a building from a stereo camera

4 Sensory Fusion

The most potential learn-on-the-fly approach is sensory fusion. Modern electronic systems such as drones and mobile phones carry many sensors: cameras, microphones, motion sensors, magnetic field sensors, GPS, WiFi, Bluetooth, cellular data, proxy distance sensors, near infrared sensors, and so on. In contrast with prevailing machine learning methods such as Convolutional Neural Networks (CNN), which require massive historical training data, learn-on-the-fly focuses on real-time lateral sensory data fusion to reveal patterns. For example, fusing a laser distance sensor data with an inertial motion unit (IMU) sensor data can enable activity recognition of firefighters with a

Decision Tree [18]. Adding more sensory dimensions increases the confidence of pattern recognition. It also improves human-machine teaming in the field of humanitarian assistance and disaster relief (HADR) tasks. For example, thermal imaging helps to detect humans and vehicles, but it has relatively low resolution compared to the visible channel. Superimposing edges on objects would assist humans and machines to identify and track the objects. Figure 5 shows screenshots of a drone video and the thermal image with edge enhancement from the visible channel.

Fig. 5. The drone image (left) and thermal image (right) of Paspardo with edge enhancement

5 Sensory Inhibition

In contrast to sensory fusion, sensory inhibition prioritizes the sensory channels in order to reduce the computational burden [6]. Sensory inhibition is also referred to as "lateral inhibition [19]," which is common in nature. For example, in neurobiology, lateral inhibition disables the spreading of action potentials from excited neurons to neighboring neurons in the lateral direction in order to create a contrast in stimulation. This happens to visual, tactile, auditory, and olfactory processing as well. For example, we do not taste our own saliva and we do not hear the sound of our jaw moving while eating. Artificial lateral inhibition has been incorporated into vision chips, hearing aids, and optical mice. Typical sensory inhibition is implemented by thresholding, delaying, and adapting. In our case, given multiple sensory channels, we find the channel that has the most contrast with the minimal processing load. Figure 6 shows a depth map and thermal image of two men laid on the floor. The thermal image shows more temperature contrast than the depth map shows distance. Therefore, to detect the human body in this case, thermal imaging would be easier. However, this imaging preference is relative and dynamic. If the person were to stand up, or if the floor temperature were as warm as the human body, then the figure-background contrast relationship would change. Figure 7 shows a color image and a depth map of two men on stairs. The depth map appears to have advantages in human detection, gesture recognition, and spatial relationship estimation when compared to a color image. Adaptation is a form of inhibition.

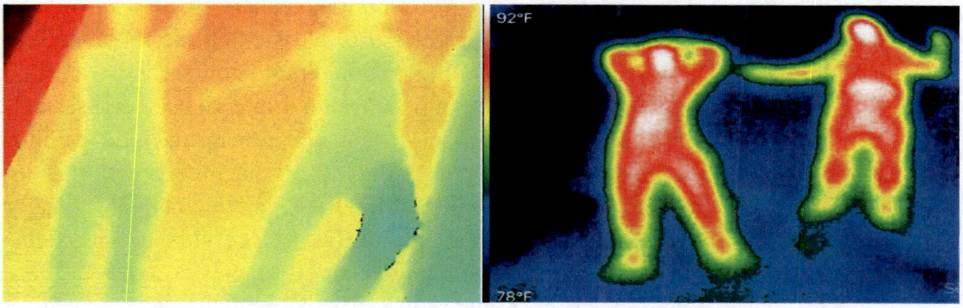

Fig. 6. The depth map (left) and thermal image (right) of men laid on the floor

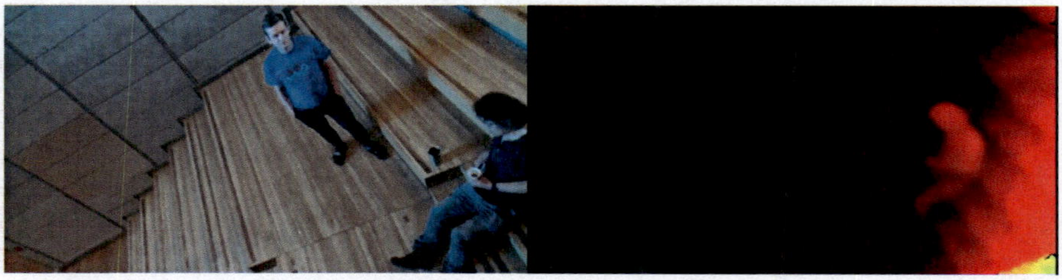

Fig. 7. The aerial color image (left) and depth map (right) of men on stairs

6 Spontaneous Alternation Behavior (SAB)

Creatures in nature commonly learn on-the-fly to adapt to changing environments. One instinctual behavior is randomization in order to search for alternative foraging paths or to avoid collision situations. When an ant gets lost, it will randomly wander until it hits a trail marked with pheromones. This pattern occurs in tests with many different animals. It is called *spontaneous alternation behavior* (SAB) [7]. Spontaneous alternation of paths for an autonomous robot, a search engine, or a problem-solving algorithm can help to explore new areas and avoid deadlock situations. Spontaneous alternation is also a primitive strategy for collision recovery. Collisions can be found in many modern electronic systems in various fields, from autonomous driving vehicles to data communication protocols. There is a variation of the SAB strategy for collision recovery. When a collision occurs, the system spontaneously switches to different sensors or channels, or the system waits for random intervals and reconnects. The "back down" and reconnect process is similar to SAB, which solves the problem of deadlock. SAB is necessary for missions involving the search for and tracking of survivors for humanitarian assistance and disaster relief (HADR). This is true especially in cases where communication breaks down, the system collapses or runs into a deadlock, or when deep, extended searches for victims in missing spots is required.

7 Summary

In this study, we explore the biologically-inspired Learn-On-The-Fly (LOTF) method that actively learns and discovers patterns with improvisation and sensory intelligence, including pheromone trails, structure from motion, sensory fusion, sensory inhibition, and spontaneous alternation. LOTF is related to classic "online modeling" or "adaptive modeling" methods. However, it aims to solve more comprehensive, ill-structured problems such as human activity recognition from a drone video in a disaster scenario. LOTF helps to build explainable AI models that enable human-machine teaming, including visual representations and visual reasoning, toward machine vision. It is anticipated that LOTF will have an impact on Artificial Intelligence, video analytics for searching and tracking survivors' activities for humanitarian assistance and disaster relief (HADR), field augmented reality, and field robotic swarms.

LOTF is an evolving approach that moves away from data-centric to sensor-centric, from rigid to adaptive, from unexplainable to explainable, from numeric to intuitive, and from curve-fitting to semantic reasoning. Our challenges include how can we scale up the system? How will we implement sensory adaptation as inhibition? Finally, how do we achieve a balance between the flexibility and efficiency of the algorithms?

Acknowledgement. The author would like to thank Sean Hackett and Florian Alber for data collection and prototyping, Professor Mel Siegel for his discussions and references on sensors and sensing, and Dennis A. Fortner for his organization. This study is in part sponsored by NIST PSCR /PSIA program and Northrop Grumman Corporation. The author is grateful to Program Managers Jeb Benson, Scott Ledgewood, Neta Ezer, Justin King, Erin A. Cherry, Isidoros Doxas, Donald D. Steiner, Paul Conoval, and Jason B. Clark for discussions, reviews and advice.

References

1. Wikipedia. On the fly. captured in 2020
2. Richman, C., Dember, W.N.: Spontaneous alternation behavior in animals a review. Curr. Psychol. Res. Rev. **5**(4), 358–391 (1986)
3. Hull, C.L.: Principles of Behavior. Appleton-Century, New York (1943)
4. Hughes, R.N.: Turn alternation in woodlice. Anim. Behav. **15**, 282–286 (1967)
5. DARPA Grand Challenge: (2016) https://en.wikipedia.org/wiki/DARPA_Grand_Challenge
6. von Békésy, G.: Sensory Inhibition. Princeton University Press (1967)
7. Cai, Y.: Instinctive Computing. Springer, London (2016)
8. Wigglesworth,V.B.: Insect Hormones. W.H. Freeman and Company (1970)
9. Cai, Y.: Ambient Diagnostics. CRC Press (2014 and 2019)
10. Horn, B.K.P, Schunck, B.G.: Determining Optical Flow. Artificial Intelligence, vol 17, pp. 185–203, Manuscript available on MIT server (1981)
11. Photogrammetry: (2020) https://en.wikipedia.org/wiki/Photogrammetry
12. OPENCV, Basic concept of the homography explained with code: (2020) https://docs.opencv.org/master/d9/dab/tutorial_homography.html
13. WikiPedia, Structure from Motion https://en.wikipedia.org/wiki/Structure_from_motion
14. Rowe, D.: Distinctive image features from scale-invariant keypoints. Int. J. Comput. Vision **60**(2), 91–110 (2004)
15. SURF: (2020) https://en.wikipedia.org/wiki/Speeded_up_robust_features
16. FAST Corner detection: (2020) https://opencv-python-tutroals.readthedocs.io/en/latest/py_tutorials/py_feature2d/py_fast/py_fast.html

17. SLAM, WikiPedia: (2020) https://en.wikipedia.org/wiki/Simultaneous_localization_and_mapping
18. Hackett, S., Cai, Y., Siegel, M.: Activity recognition from firefighter's Helmet. In: Proceedings of CISP-BMEI, Huaqiao, China (2019)
19. Lateral Inhibition, WikiPedia: (2020) https://en.wikipedia.org/wiki/Lateral_inhibition

Boeing 737 MAX: Expectation of Human Capability in Highly Automated Systems

Zachary Spielman(✉) and Katya Le Blanc

Idaho National Laboratory, 1955 N. Fremont Avenue, Idaho Falls, ID 83415, USA
{zachary.spielman,katya.leblanc}@inl.gov

Abstract. The tragedy of Lion Air Flight 610 and Ethiopian airlines flight 302 offer important insights into the dangers of advancing automation without a complete understanding of the context in which the system will be used. This paper recounts the story of Lion Air flight 610, and provides a brief analysis of the organizational, market, and cultural factors that resulted in the crash. The analysis highlights how a series of failures in each of the factors can compound to result in catastrophic consequences. The lessons learned are applied to the nuclear industry to describe how similar failures could manifest in the nuclear industry if designers, regulators, and operators don't carefully consider similar factors.

Keywords: Human Factors · Human-systems integration · Automation · Boeing 737

1 Introduction

In 2018, one airplane crashed killing all on board. Five months later, the same model aircraft crashed succumbing to the same symptoms as the first.

Flight is historically a safe mode of travel compared to other options. Pilots are rigorously trained in high-fidelity flight simulators and hand-selected to pilot commercial airliners. Aviation and cock-pit design is considered by many the birthplace of Human Factors research; the two fields working hand in hand since World War II. Regulatory standards are mature and enforced. Current aircrafts also have redundancies and automated safety systems to account for machine and pilot error. So, how is it a worst-case scenario occurred during not one, but two separate flights?

The piloting crew of both aircrafts failed to correctly respond to a violent run-away-trim or, at least, a scenario that appeared like one. In fact, an automated system called "Maneuvering Characteristic Augmentation System" (MCAS) used information from a single faulty sensor pitching the front of the plane towards the ground over and over again. The MCAS was supposed to prevent the plane from stalling due to aerodynamic issues with new engines installed on the 737 wings. The faulty sensor told the MCAS the plane was in danger of stalling initiating automatic nose down actions forcing the two flights to ultimately crash.

M. Zallio (Ed.): AHFE 2020, AISC 1210, pp. 64–70, 2021.
https://doi.org/10.1007/978-3-030-51758-8_9

These two catastrophes led to grounding every single Boeing 737-MAX plane in the world. This human factor analysis found that Lion Air, Boeing, and the FAA all intended to maintain multiple safety nets to prevent this type of catastrophe. However, due to organizational, cultural, and market factors, each of them made assumptions that rendered these safety nets ineffective in the context that the Lion Air and Ethiopian flights operated in. The result was an inadequate safety system relying on a malfunctioning part in the hands of poorly trained pilots.

This event occurred at a time the aviation industry was booming. Advances in automation and fly-by-wire technology allowed airlines in less regulated countries to grow rapidly. Due to abundant procedures, checklists, and cockpit safety systems training became a lower priority. This culture shift was not accounted for by Boeing who abided a philosophy that "if all else fails, the pilot knows what to do." Other industries developing more "accessible" technology, such as Nuclear power generation, can learn a valuable lesson from this tragedy. As technology becomes more accessible, designers and operators must be aware of the risk that cultural differences may lead to invalid assumption about the conditions a plant will be operated in.

2 Lion Air Flight 610

Almost immediately after take-off air traffic control should have recognized the flight was in distress. The first officer in the cockpit kept asking air traffic control to confirm their altitude and speed despite redundant cockpit instrumentation. Then the plane made a couple violent dives. The first officer requested clearance to level off at 5000 feet as a result of a "Flight Control Problem." The aircraft shakily climbed to 5000 feet while receiving several new headings. As the plane banked to follow each new heading it became clearer the pilots were having difficulty keeping the plane steady. This behavior was strange but, when asked prompted the pilots never reported an emergency on-board.

The aircraft involved in flight 610 had faulty sensors. The faulty sensors meant the captain and first officer were receiving different information and cockpit alarms were sounding to warn the flight of this issue. One sensor in particular, the angle-of-attack sensor on the captain's side was most important. A new fail-safe software had been installed on this aircraft that relied on information from this single sensor to determine if it needed to correct the planes attitude. This single sensor was misreading the planes pitch as 21° higher than what was true. Due to this miscalculation the software interceded and attempted to correct the aircrafts pitch downward. Doing so meant forcing a level plane to pitch downward towards the sea. Had this been the only problem aboard flight 610 the pilots may have known what to do. The cockpit however was chaotic and distracted.

The pilots of flight 610 were overwhelmed. Cockpit alarms began sounding at the moment of take-off until the end of the eleven-minute flight [1]. During the 11-min flight the first officer received 11 communications and 11 new headings from air traffic control (totaling 22 interactions). At the same time the captain was instructing him to perform "memory items" for airspeed unreliable situations; a memorized checklist preceding the paper procedure (which he was also looking for). The captain had a constant "stick shaker" alarm. An alarm that vibrates the yoke (a planes steering wheel) to indicate a possible stall event. Four minutes into the flight the captain began battling the software adjusting elevator trim forcing the plane downward. The MCAS forced the plane nose down while speed, altitude, angle-of-attack, ground proximity, and other alarms were already sounding. Keeping the plane level was further complicated by adhering to new heading coordinates each minute requiring more complex maneuvers. Making matters worse, in the final minutes, additional flight crew began entering the cabin and talking to the pilots trying to diagnose the situation. The combination of everything overwhelmed the pilots and the plane was lost.

It likely appeared to outside observers the plane had a run-away trim scenario; a case where the automatic trim adjustment on the tail of the plane malfunctioned causing the plane nose to dive or climb unexpectedly. The simple solution is using the trim "CUT OUT" switches disabling the automated fly-by-wire control. Doing so would have ended the violent dives. Although the intense pitch adjustments were due to MCAS and not a run-away trim this would still have worked in flight 610. The trim cut out switches were the ad hoc solution to disable the new MCAS software designed to prevent stall events.

Run-away trim is a common training scenario. Perhaps, had it been the only problem, the first-time captain would have recognized it. In fact, the flight crew that last used this aircraft had recognized it, almost. Three days earlier a similar situation occurred with a couple key differences. During pre-flight briefings the pilots were informed as mandated that the angle-of-attack sensor on the captain's side had been replaced. Also, they had a third pilot riding in the cabin, off duty, deadheading back to Jakarta. During take-off the pilots received many of the same alarms as flight 610 which included disagreement alarms but, most importantly, a large 21-degree discrepancy between the captain's angle-of-attack sensor and the first officers. Shortly after take-off the plane started violently pitching downward. The captain and first officer reacted in the same way as flight 610. Watching the events unfold from the back seat the passenger pilot suggested using the trim cut out switches. The plane stopped its violent behavior. The flight made it successfully to Jakarta even though sensor disagreements persisted. Unfortunately, no report of the issue, the faulty sensor or the solution was made or communicated to the novice flight crew of flight 610.

Flight 610 was the captain's first commanding flight and his first officer was less experienced. Both had recently graduated Lion Air's in-house training program. Viewing the pilot's performance and decision making in isolation it seems their incompetence was the obvious culprit. However, the situation they experience, that occurred two other times in very similar fashion, points to a more systemic issue.

3 Aviation Culture

Advancements in aircraft technology tend to improve either aircraft safety or flight efficiency. Adding fail-safe systems or improved human-system communication can minimize the error consequences and provide pilots with decision making tools and advanced plane diagnostics to assess the health of the plane and validity of its sensors. Early aviation philosophy viewed the cockpit as a secondary to pilot authority. Basically, help the pilot as much as possible but, in the end, the pilot will know what to do. As a result, rigorous training and education is required of aspiring pilots before handing them the controls. But cockpit technology, especially fly-by-wire technology, sprouted a new philosophy developed in the 1980's; keep the plane safe from pilot error. The new culture swaps the role of pilot and cockpit. Airbus, a manufacturing competitor to Boeing, is known for adopting this view in the design of their aircrafts. It was this shift where developing airlines like Lion Air saw advanced cockpit technology as an opportunity.

3.1 Lion Air Culture

Lion Air is an economy airline in Indonesia. Their goal is to fill seats and ship passengers. Lion Air achieved this by dramatically reducing pilot training requirements and insufficient quality control on parts and maintenance, ultimately at a cost to safety. Indonesia had little regulatory standards to hold airlines accountable and Lion Air took advantage of that. They maintained their own pilot academy guaranteeing every graduate a job as a copilot and promoting most to captain within four years. In contrast, U.S. based ATP Career pilot training graduates about 80 percent of enrollees none of which are guaranteed a position with an airline and the US Air Force Academy graduates only 50 percent of candidates. Being promoted to captain in four years is also accelerated with US sources stating 10–15 years in larger airlines [3]. Remarkably, this worked. For most flights, once planes were in the air, pilots could make it to their destination safely.

Lion Air pilots were able to achieve this through rote memorization and support from procedure, checklists, and cockpit support software. However, their training did not train them to handle unfamiliar conditions such as those experienced on the day of the crash. Their flight experience was curtailed due to crowded training simulators. Typical flight simulators had two pilots and an instructor. Their simulator training often had 7 people attending, two acting as pilot, one instructor and four watching the pilots in training from behind the instructor [4]. Lion Air pilots recognized flight hazards by their order in the simulator and not by their symptoms. Exacerbated by poor maintenance; conflicting sensors, indications and cockpit warnings were commonplace, often appearing without explanation. On the ground, mistakes happened all the time, maintenance practices were poor and issue tracking and reporting even worse. The safety impact of Lion Air's culture was recognized worldwide. The US and UK, from 2007 to 2016, banned Lion Air from their airports citing a risk to their citizens safety. The airlines philosophy might have reflected the idea that these advanced planes are protected from pilot error. It is perhaps why they made one of the largest orders of the newest Boeing model in 2016.

3.2 Boeing Culture

Boeing is known for their "pilot first" philosophy; that the cockpit is a tool to support the pilot. They support this philosophy by working to raise the standards for pilot training in the airlines they sell too. Chinese airlines are benefactors of this effort moving their pilots from rote memorization and reliance on step-by-step procedures to pilots with airmanship their safety record improved. Boeing attempted the same thing in Indonesia, but they suffered from a lack of regulatory support and their pilot training was left unchanged [4]. Despite knowledge that not all airline pilots are created equal Boeing still relied on this philosophy during the development of the 737-MAX. Boeing made a series of alterations to their popular 737–800 aircraft and patched shortcomings with software. They assumed the pilots using their planes would not be impacted so they decided not to release any information about the new software to any of the airlines they sold too.

It was Airbus that announced the Airbus-neo (New Engine Option) with improved fuel efficiency. The announcement forced Boeing to quickly respond else lose out on a 35-billion-dollar market opportunity. Airbus and Boeing have a history of competition often releasing new planes in the same model year that claim further distances using less fuel or carrying more weight [4]. Boeing had to announce a competing design fast. The Boeing 737-MAX design was announced in mid-2011, 9 months after the Airbus-NEO (Fig. 1).

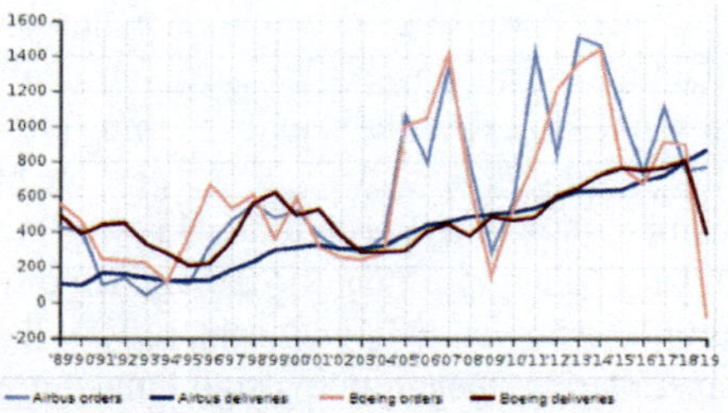

Fig. 1. Annual net orders on aircraft deliveries by airbus and Boeing Commercial airplanes. Graph created by Wikipedia but sourced from [5] and [6]

To be successful, at least four factors had to be met. The new design had to be on the Boeing 737 model. The same "Type Certificate" had to be used. The plane had to improve efficiency. And above all, it had to be done fast.

4 FAA Certification

Boeing has a few *Type* builds such as the 737, 747, 787, and 777. They range in size, range, and passenger capacity. The aircraft type is important. It is expensive to certify a pilot on an aircraft type but inexpensive to keep them certified. Furthermore, it is very expensive for manufactures to get regulatory approval for a new aircraft *Type* but much less expensive to make slight modifications to existing types. Therefor it is advantageous to both airlines and manufacturers to build and purchase the same *type* aircrafts. Boeing chose to modify the 737–800 to best compete with the Airbus-NEO. Doing so meant installing new engines. Mounting the new (larger), engines in the same place caused complications with the landing gear, resulting in a new mounting location. The new mounting location mixed with the new engine power had a negative impact on flight handling characteristics at high angle-of-attacks. To compensate the negative handling characteristic Boeing developed new software designed to counter the stall risk at high angle of attack called "Maneuvering Characteristic Augmentation System" or MCAS.

Boeing performs much of the certification process in house. The Federal Aviation Administration (FAA) then reviews the work for final judgement. This method is done to keep FAA budget low and personnel demand manageable (CITE). Historically this method has been successful.

Boeing rationalized much of the MCAS viability by falling back on their legacy credo, trusting the pilot to intercede if something went wrong. If failed, the MCAS looked and felt like a common "Runaway trim" situation that all pilots are trained to detect and correct. Even the same correction was required; flip the automated stabilizer trim cut out switches. The cut outs would deactivate the MCAS returning all control to the pilot. Because a failing MCAS replicated such a common event, pilots were never informed a new automated system existed on the plane. Also, the MCAS was considered a last resort measure. The stall levels required to activate it were far outside a range any reasonable pilot would every find themselves in. Boeing did perform some simulator tests to verify how a pilot react to "uncommanded MCAS Activation." However, no test or simulation replicated a failure in sensors leading to inappropriate MCAS activation.

The FAA reviewed MCAS classifying its failure as "Hazardous." That classification requires two levels of redundancy and a failure chance of 1 in 10 million. At the time of review the MCAS specification stated it would alter trim ~.06°. The possibility of the MCAS failing multiple instances in short succession was never reviewed. Somehow, between the review and implementation, the MCAS relied on a single sensor, the captain's angle-of-attack sensor, to activate. When activated it could adjust trim 2.5°; four times the amount stated during review. Also, in both recorded crashes, the MCAS repeatedly failed continually adjusting elevator angle pitching the plane further downward [8].

5 Lessons Learned

The capability of automation is constantly increasing. The human role in controlling a system evolves as automation capabilities take over more complex tasks. Lion Air is a dramatic example on this overreliance on automation. Instead of evolving the role of the pilot with a full understanding of the limitations of the automation, they relaxed pilot qualifications assuming automation would cover the difference. As more safety nets were violated (operator training, maintenance practices, reporting cultures) they increased the likelihood of risk resulting in catastrophic consequence. Accelerating the process was Boeing's assumptions that their aircrafts would be handled by pilots trained to their standards and airlines operating culture matching their expectation. The combination led to disaster. As other industries approach this nexus in automation and human expectation, they can learn what not to do from the Boeing 737-MAX case.

The nuclear industry is advancing automation, in plant design, and, perhaps most importantly, in application. Smaller, highly automated plants may be delivered to communities that differ greatly from the traditional operator. The new designs are inherently safe, but the positive public perception of nuclear energy is more frail than the perception of the aviation industry. Lessons from the Boeing 737-MAX are useful to understanding how old expectations are liabilities to progressing the industry.

It is important that new design and concept of operations are vetted within the same conditions they will be operated. Companies should take initiative to train and promote safe regulation where their customers will operate. Updating automation to increase safety and efficiency is necessary. Also, necessary is evaluating how new automation impacts the role (or perception thereof) those interacting with the system. If the operators must take on new roles, are they supported to a safe extent?

Nuclear power generation technology is becoming smaller, cheaper, and more accessible. The new technology may provide power to remote, powerless, or disaster-stricken parts of the world. Regulatory bodies are not accustomed to reviewing and validating the new reactor designs. Also, operational safety is validated using highly knowledgeable, trained operators. Real-world applications might not reflect the conditions these plants are tested in. The result from this investigation is a cautionary tale for industries outside aviation.

References

1. Tjahjono, S.: Komite Nasional Keselamatan Transportasi Republic Of Indonesia. Tanjung Karawang, West Java. Komite Nasional Keselamatan Transportasi (2019)
2. Sumwalt, R.L., Landsberg, B., Homendy, J.: Assumptions Used in the Safety Assessment Process and the Effects of Multiple Alerts and Indications on Pilot Performance. District of Columbia: National Transportation Safety Board (2019)
3. How Long Does It Take To Become A Captain? (n.d.). https://www.flightdeckfriend.com/how-long-does-it-take-to-become-a-captain/. Accessed 13 Feb 2020
4. Langewiesche, W.: What Really Brought Down the Boeing 737 Max?, 18 September 2019. https://www.nytimes.com/2019/09/18/magazine/boeing-737-max-crashes.html
5. Mish: Boeing 737 Max Major Design Flaws, Not a Software Failure, 17 March 2019. https://moneymaven.io/mishtalk/economics/boeing-737-max-major-design-flaws-not-a-software-failure-rVjJZBVzZkuZLkDJn3Jy8A/
6. Boeing Orders and Deliveries. (n.d.). http://www.boeing.com/commercial/#/orders-deliverie. Accessed 13 Feb 2020
7. Airbus Orders and deliveries (n.d.). https://www.airbus.com/aircraft/market/orders-deliveries.html. Accessed 13 Feb 2020
8. Gates, D.: Flawed analysis, failed oversight: How Boeing, FAA certified the suspect 737 MAX flight control system, 21 March 2019. https://www.seattletimes.com/business/boeing-aerospace/failed-certification-faa-missed-safety-issues-in-the-737-max-system-implicated-in-the-lion-air-crash/

Reducing Drone Incidents by Incorporating Human Factors in the Drone and Drone Pilot Accreditation Process

Daniela Doroftei(✉), Geert De Cubber, and Hans De Smet

Royal Military Academy, Avenue De La Renaissance 30, 1000 Brussels, Belgium
{Daniela.Doroftei,Geert.De.Cubber,Hans.DeSmet}@rma.ac.be

Abstract. Considering the ever-increasing use of drones in a plentitude of application areas, the risk is that also an ever-increasing number of drone incidents would be observed. Research has shown that a large majority of all incidents with drones is due not to technological, but to human error. An advanced risk-reduction methodology, focusing on the human element, is thus required in order to allow for the safe use of drones. In this paper, we therefore introduce a novel concept to provide a qualitative and quantitative assessment of the performance of the drone operator. The proposed methodology is based on one hand upon the development of standardized test methodologies and on the other hand on human performance modeling of the drone operators in a highly realistic simulation environment.

Keywords: Human factors · Drones · Performance analysis

1 Introduction, Motivation and Scope of Work

The number of small drones and drone operations is expanding and proliferating tremendously. However, there is a problem. Drones crash. Often [1]. When they do, international studies [2] show that around 80% of the drone crashes can be related to human factors. Combining these two facts, it is clear that - if we want to avoid a massive number of drone incidents in the future – it is required to develop a strategy to incorporate human factors in the drone deployment process and the training of drone pilots.

Pilots for regular aircraft or for larger (typically military) drones generally follow extensive simulator training before engaging in any real flight. However, for small rotorcraft, this is much less the case, because it is very difficult to convey a realistic representation to the human sensory system. Both for fixed wing and rotary wing drones, the main problem with current simulator-based pilot training programs is that they are limited to simplistic scenarios (typically flying predefined patterns and practicing take-off and landing operations), without providing much qualitative feedback to the trainee or the supervising entity.

In response to these identified shortcomings, we are developing a drone operator performance assessment tool, which uses a realistic environment and realistic operational conditions to measure the performance of the drone operator, both in a qualitative

M. Zallio (Ed.): AHFE 2020, AISC 1210, pp. 71–77, 2021.
https://doi.org/10.1007/978-3-030-51758-8_10

and quantitative manner. These metrics can then be used by training responsibles to adapt/adjust the theoretical and practical training courses for drone pilots, such that the curriculum (both the practical and the theoretical courses) can be iteratively optimized to best fit the needs.

An important aspect of any qualification assessment procedure is the definition of the test methodologies and of the test scenarios. Within the subject of drone pilot training, these test scenarios are currently most often very limited to simple take-off & landing operations and of following simple patterns in the air. For pilots working in the security sector (military, police, firefighters, civil protection, …) in tough operating conditions, these highly simplistic scenarios are hardly relevant. Therefore, we also propose a set of standard test methods specifically geared towards the training of drone operators in the security sector.

2 Previous Work and Main Contributions

Drone operator human performance models have been first developed by the US Air Force in [3], focusing on operations with large military drones, navigated by a crew. These military-oriented operator performance modeling approaches tend to focus on operator workload analysis for optimizing the crew composition, which is less relevant for micro-UAS systems. Bertuccelli et al. [4] proposed a new formulation for a single operator performing a search mission with multiple drones in a time-constrained environment. Wu et al. [5] expanded on this idea by proposing a multi-operator multi-drone operator model.

The main criticism with respect to these approaches is that they focus heavily on aspects such as attention and fatigue modeling and neglect other aspects that are paramount for operations in the security sector such as mission stress, enemy counter-measures, varying operator skill levels, etc.

In response to these identified shortcomings, we propose a holistic drone operator performance model, targeted towards drone operators in the security sector, taking into consideration identified parameters that are critical towards these end users.

Efforts to integrate drones into standard operating procedures and into the operational toolbox of security operatives would benefit from quantitative evaluations of individual aircraft capabilities and associated remote pilot proficiencies. The National Institute of Standards and Technology (NIST) is leading an international effort to develop the measurements and standards infrastructure necessary to quantitatively evaluate such aircraft and pilots in the framework of urban search and rescue (USAR) operations [6]. The resulting standard test methods enable any user to generate statistically-significant performance data to evaluate airworthiness, maneuvering, sensing, payload functionality, etc. While extremely valuable, these standard test methodologies developed by NIST are heavily focused on USAR operations and not generically useable for all type of security operations.

Therefore, we propose a set of standardized test methodologies for security operations, based upon the existing NIST framework for USAR operations.

3 Conceptual Overview of the Methodology

In order to assess the relationship between human factors and the human operator performance, we followed a user-centered design [7] to come to the methodology which is graphically depicted on Fig. 1 and which can be summarized as follows:

1. Identify which human factors could potentially impact the performance of drone pilots, via a set of interviews with experienced drone operators. From this set of interviews, the following human factors were discerned as most important:

Table 1. Most important human factors impacting drone operator performance.

Human factor	Importance level (0–100%)
Task difficulty	89%
Pilot position	83%
Pilot stress	83%
Pilot fatigue	83%
Pressure	83%
Pilot subjected to water or humidity	83%
Pilot subjected to temperature changes	78%
Information location & organization & formatting & brightness of the controller display	78%
Task complexity	78%
Task duration	78%
Pilot subjected to low quality breathing air	72%
Pilot subjected to small body clearance	72%
Ease-of-use of the controller	72%
Pilot subjected to noise/dust/vibrations	67%
Task type	67%

It should be noticed that these scores were given by expert operatives. We inquired for many more potentially influencing factors and some (e.g. distraction) scored suspiciously low, so they did not make it to the list of important factors of Table 1. Notwithstanding this, we also test against these factors in the evaluation process.

Each of these identified parameters is re-identified with the test subjects (drone pilots) during an intake questionnaire to assess the state of the pilot when she or he starts the simulation exercise.

2. Identify which operational scenarios and environmental conditions could potentially impact the performance of drone pilots, via a set of interviews with experienced drone operators working in the security sector. From this set of interviews,

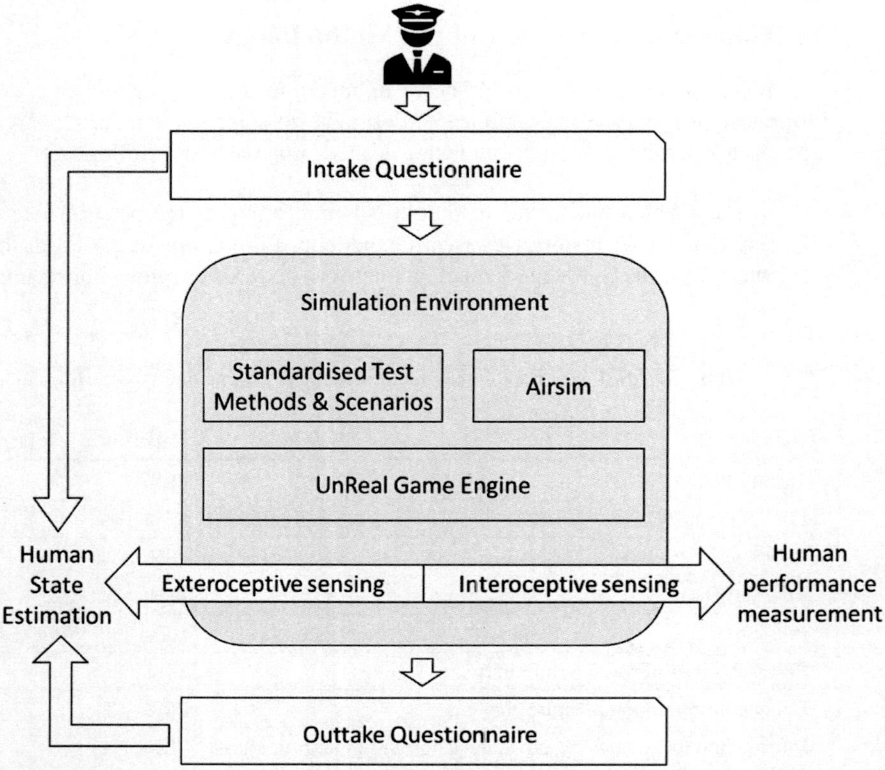

Fig. 1. Schematic overview of the test procedure where the pilots are subjected to. After taking an intake survey, the pilots have to perform a complex mission in a simulation environment. While doing this, their performance parameters and physiological state are measured. After completing the mission, they perform an outtake survey.

a set of standard operational scenarios were compiled that cater to the needs of as many end-users (drone operatives in the security sector) as possible. These scenarios consider complex target observation & identification missions in urban and rural environments.

3. Development of a simulation environment for complex drone operations. Driven by the innovation in the field of game development and the increasing graphical processing power of computers, current simulator engines provide a very realistic environmental representation, and the integration with virtual and augmented reality systems allows to increase the level of realism even further. All this means that the visual quality provided by the existing engines is generally very high. However, they also have some important disadvantages, as most existing simulator engines are closed solutions and thus provide no possibility to integrate added functionalities. Therefore, we use the Microsoft AirSim simulation engine [8], which is an open-source simulator for RPAS built on the Unreal Engine. This simulation environment is completely open and customizable, which enables us to incorporate the standard

test scenarios, multiple customizable drones and to quantitatively measure the performance of the pilots on-line while executing the mission. Next to this interoceptive sensing of the human physiological state, we also plan to use (in a later stage) exteroceptive sensing of the human physiological state by a camera system targeted at the pilot, estimating fatigue etc.

We make use of a human-machine interface with a curved monitor, and not of a virtual reality interface, as may be expected. While the simulation engine supports virtual reality and we have the equipment available, we have especially opted for not making use of a virtual reality interface for two reasons:

1. We want to avoid measuring the side-effects of virtual embodiment, where some pilots may be subject to.
2. Virtual reality would obstruct the use of exteroceptive sensing tools for measuring the physiological state of the pilot during the test.

At this moment, we work on two scenarios within the simulation:

- Stealthy detection and observation of enemy forces in a rural environment
- Management of a hostage situation in an urban environment

In each of these scenarios, the pilots are confronted with large-scale dynamic environments, changing weather conditions and time pressure in order to deliver quality data in a minimal amount of time, which are all factors that can induce human errors that can dramatically impact the performance.

4. After completing the mission, the test subjects will be asked to fill in another questionnaire in order to assess their physiological state, as well as to assess any differences with respect to the moment of performing the intake survey.
5. At the end of this procedure, this means that we have the following data at our disposal:

    a. Human factors & human physiological state prior to beginning the mission (through the intake questionnaire)
    b. Human factors & human physiological state during the mission (through the exteroceptive sensing, though this is still under development)
    c. Human factors & human physiological state after completing the mission (through the outtake questionnaire)
    d. Human performance data as quantitatively measured using the simulation engine (interoceptive sensing), which is directly usable for pilot performance assessment.

Given enough test subjects, this enables us to set up a mathematical model between on one hand the human factors and the human physiological state and on the other hand the human performance. This model enables us to predict human performance given a certain input state.

In the next section, we will explain how such a model can also be effectively used in the drone pilot accreditation process and the drone certification process.

4 Incorporation of the Human Performance Model in the Drone and Drone Pilot Accreditation Process

In manned aviation, there exist extremely strict procedures for pilot accreditation and aircraft airworthiness certification. For small unmanned aircraft, however, the rules are less tight and also less harmonized globally. In the European Union, a risk-based approach [9] is followed, where tighter and tighter rules are imposed (both for the pilot license as for the aircraft airworthiness assessment) with increasing risk associated to the drone operation to be performed. A crucial point is thus to assess the risk to a drone operation, which is dependent of the scenarios that are going to be performed and that are written down in the operational handbook. Therefore, a set of standard scenarios are defined and in order to get a permission to fly, the performance of drone pilots and drones for a specific scenario needs to be assessed. This concept of operation for the accreditation has two pitfalls that our work tries to address:

1. **The drone pilot accreditation process** happens once, once a year or once every few years. However, we know that a varying physiological state of the pilot on the date of the flight may impact the performance drastically. Using our human performance model, we can predict – given a certain physiological input state – what would be the flight performance of the human operator. As such, a much more fine-grained case-based accreditation is possible, which is specifically useful for stressful operations, such as is often the case in the security sector.
2. **The drone accreditation process** is not pilot-agnostic. Indeed, when a new drone is tested in a real or simulation environment, it is controlled by a human pilot. Each pilot is of course different, and the performance of each pilot will differ, which will have its impact on the evaluation of the drone under investigation. Using our human performance model, we can create a generic computer pilot that is able to perform a flight operation to test a drone system without any influence of a human pilot. This would provide a much fairer assessment for the accreditation of drones and a valuable extra metric to be taken into consideration for the airworthiness assessment.

5 Conclusions and Future Work

In this paper, we have presented a methodology for the qualitative and quantitative assessment of drone pilots, which can then be used as a tool for improving the training curriculum for these drone pilots. The methodology is based upon a virtual training environment and a set of standard test methods. Importantly, the proposed methodology also enables the development of a human performance model, interlinking the human factors and physiological state on one hand and the human performance on the other hand. This is a crucial tool, as it would not only teach us the relationship between these parameters, but it would in a later stage also support completely pilot-agnostic qualitative [10] and quantitative [11] evaluation of drones and drone pilots.

The proposed methodology is currently under validation with real drone pilots. The first feedback shows that the users appreciate the level of visual detail and realism; however, they do indicate that in the control of the vehicle the sense of realism is missing, which falsifies the feedback we get from the test results. This is certainly a point that needs to be improved further.

Obviously, the presented methodology is not a final product yet and requires still a lot of work. Currently, we are working on improving the simulation engine in order to better deliver the level of realism the end users request in terms of controls for the vehicle. Once this is done, we will launch a large test campaign with dozens of pilots, which will not only allow us to qualitative and qualitative assess the performance of these pilots, but also to build up the human performance model, as discussed in Sect. 3. This human performance model will be used later on one hand as a reference for drone pilot performance testing and on the other hand for the assisting with the accreditation of new drone designs, as it would allow to eliminate the human pilot from the test process.

References

1. Chow, E., Cuadra, A., Whitlock, C.: Hazard Above: Drone Crash Database - Fallen from the skies. The Washington Post, Washington, D.C., 19 January 2016 (2016)
2. Shively, J.: Human performance issues in remotely piloted aircraft systems. In: ICAO Conference on Remotely Piloted or Piloted: Sharing One Aerospace System (2015)
3. Deutsch, S.: UAV operator human performance models. BBN Report No. 8460. Technical report (2006)
4. Bertuccelli, L.F., Beckers, N.W.M., Cummings, M.L.: Developing operator models for UAV search scheduling. In: Proceeding of AIAA Guidance, Navigation, and Control Conference, Toronto, Canada (2010)
5. Wu, Y., Huang, Z., Li, Y., Wang, Z.: Modeling multioperator multi-UAV operator attention allocation problem based on maximizing the global reward. In: Mathematical Problems in Engineering (2016)
6. Jacoff, A., Saidi, K.: Measuring and comparing small unmanned aircraft system capabilities and pilot proficiency using standard test methods. NIST report, April 2017
7. Doroftei, D., De Cubber, G., Wagemans, R., Matos, A., Silva, E., Lobo, V., Cardoso, G., Chintamani, K., Govindaraj, S., Gancet, J., Serrano, D.: User-centered design. In: De Cubber, G., Doroftei, D. (eds.) Search and rescue robotics - from theory to practice. InTech (2017)
8. Shah, S., Dey, D., Lovett, C., Kapoor, A.: AirSim: high-fidelity visual and physical simulation for autonomous vehicles. In: Field and Service Robotics, vol. 5, pp. 621–635. Springer (2017)
9. European Commission: Commission Implementing Regulation (EU) 2019/947 of 24 May 2019 on the rules and procedures for the operation of unmanned aircraft (2020)
10. Doroftei, D., Matos, A., Silva, E., Lobo, V., Wagemans, R., De Cubber, G.: Operational validation of robots for risky environments. In: Proceeding of 8th IARP Workshop on Robotics for Risky Environments, Lisbon, Portugal (2015)
11. De Cubber, G., Doroftei, D., Balta, H., Matos, A., Silva, E., Serrano, D., Govindaraj, S., Roda, R., Lobo, V., Marques, M., Wagemans, R.: Operational validation of search and rescue robots. In: De Cubber, G., Doroftei, D., (eds.) Search and Rescue Robotics - From Theory to Practice. InTech (2017)

Use a UAV System to Enhance Port Security in Unconstrained Environment

Ruobing Zhao[1], Zanbo Zhu[2], Yueqing Li[1](✉), Jing Zhang[2], and Xiao Zhang[3]

[1] Industrial Engineering Department, Lamar University, Beaumont, TX, USA
{rzhao1,yueqing.li}@lamar.edu
[2] Computer Science Department, Lamar University, Beaumont, TX, USA
{zzhu4,jzhang9}@lamar.edu
[3] College of Business, Lamar University, Beaumont, TX, USA
xzhang8@lamar.edu

Abstract. Ensuring maritime port security—a rapidly increasing concern in a post-9/11 world—presents certain operational challenges. As batteries and electric motors grow increasingly lighter and more powerful, unmanned aerial vehicles (UAVs) have been shown to be capable of enhancing a surveillance system's capabilities and mitigating its vulnerabilities. In this paper, we looked at the current role unmanned systems are playing in port security and proposed an image-based method to enhance port security. The proposed method uses UAV real-time videos to detect and identify humans via human body detection and facial recognition. Experiments evaluated the system in real-time under differing environmental, daylight, and weather conditions. Three parameters were used to test feasibility: distance, height and angle. The findings suggest UAVs as an affordable, effective tool that may greatly enhance port safety and security.

Keywords: Port security · Unmanned aerial vehicles · Human body detection · Human facial recognition

1 Introduction

Port security, a specific area of maritime security, refers to the safety and law enforcement measures that fall within the port and maritime domain and includes the protection and inspection of cargo transported through the ports from unlawful activities. Port security is crucial to national security, as maritime transport serves as a primary means of international trade and transport of goods. Proper monitoring and inspection are essential to preventing the inappropriate use of cargo containers. Port security discussions often revolve around prevalent maritime threats, such as terrorism, organized crime, smuggling, piracy, and even accidental spills and other cargo release events. For example, since October 1st, 2019, Homeland Security and its partners have recorded more than 100 smuggling attempts along the coast, according to CBP [1].

Before the September 2001 terrorism incidents, the most significant port security threats were organized crime and drug smuggling. However, since then, terrorist attacks

M. Zallio (Ed.): AHFE 2020, AISC 1210, pp. 78–84, 2021.
https://doi.org/10.1007/978-3-030-51758-8_11

have become a major issue in port-related security risks. The need for improved maritime port security has increased in recent years, since terrorists and pirates have begun using sea routes to cause greater levels of damage to society.

As the need for enhanced maritime and port security has intensified, technology developers have responded, producing next-generation security control systems with enhanced video surveillance, alarm monitoring, and other features. More recently, unmanned aerial vehicle (UAV) technology has evolved to address certain operational challenges, as standard port surveillance systems must operate in a complex environment with a high volume of merchant and fishing vessel traffic. Drones can be remotely controlled by radio/infrared communication, eliminating the need for pilots and increasing human safety. UAVs can also mitigate the inherent vulnerabilities of ground radar systems: fixed radar cannot cover "blind zones." In addition, drones possess the range and ability to deliver accurate real-time images to ground staff who can make informed, timely decisions based on that intelligence. As their batteries and electric motors grow increasingly lighter and more powerful, UAVs have been shown to be capable of enhancing a surveillance system's capabilities and mitigating its vulnerabilities. Integrating one or two UAVs into a traditional surveillance system can enhance the detection efficiency and make it possible to continuously monitor enemy and neutral targets. Exigent examples abound of this new drone technology in action. Japan's Sky Remote Company has developed Kite UAV, which is being used for land, coastline, and ocean surveying in China [2]. Abu Dhabi Ports Company (ADPC) has reported that two remote-controlled flying drones are patrolling Abu Dhabi's ports to strengthen maritime security [3].

This paper focused primarily on enhancing port security using UAV (drone) systems. The study tested the feasibility of a collaborative UAV system in an unconstrained environment through the deep learning method. Human body detection and facial recognition algorithms were used to identify humans in UAV real-time videos.

2 Methodology

The proposed system consists of three major steps: (1) system setup; (2) human body detection; and (3) human facial detection and recognition. The methodologies for these steps are explained in the following three sections.

2.1 System Setup

To detect suspicious targets, an Android application called DJI Go 4 was used to receive and transmit UAV camera data. The transmission of data from an Android device and reception of the data on the laptop PC was achieved using nginx Real-Time Messaging Protocol (RTMP). The UAV remote controller and the Android device were connected via USB cable, and the Android device was connected to the computer through the laptop's hotspot local network. The IP address of the computer from which to send the data was assigned in the Android application. The PC acted as a server and received data from the Android device.

Figure 1 illustrates the UAV-based real-time human detection and facial recognition system. The UAV was set by the DJI GO 4 App to fly through several preset waypoints at a high altitude to monitor the situation of a designated area. The video captured by the UAV onboard camera was transmitted to the Android device on the remote controller. This video was then streamed by Android device to the PC using local wireless network. Once a suspicious person was detected, a second drone flew to the area where the person was detected. This drone was controlled manually and flew at a closer range for the purpose of human facial detection and recognition. The same streaming system was applied so that the real-time video could be transferred to a PC where a facial detection and recognition program was running.

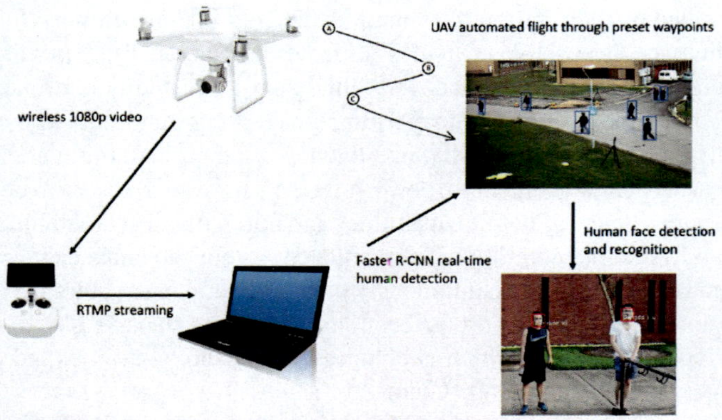

Fig. 1. UAV-based human detection and facial recognition system

2.2 Human Body Detection

Based on the experiments, deep learning-based approaches perform significantly better than traditional histograms of oriented gradients (HOG) [4] and Haar [5, 6] approaches for human detection. Among those, the Faster RCN Inception V2 COCO Model gives a much better result. The Faster R-CNN method [7, 8] was applied to process video frames on the PC for human detection in the system.

Compared with traditional image feature-based approaches [9, 10], deep neural network-based approaches provide more accurate results at cost of more computations. Most problems in early human detection methods are solved, and these fixes are introduced at the cost of more computations. However, in the presence of GPU acceleration, modern machine learning libraries can provide these improved results with comparable frame rates. The Faster RCN Inception V2 COCO Model gives a fair trade-off between accuracy and speed for GPU accelerated environments. It also provides a tight, consistent boundary around a detected person (see Fig. 2).

2.3 Human Facial Recognition

After the human body has been detected, facial recognition will be performed to match the suspicious target in the database. In this case, a deep learning-based facial recognition

Fig. 2. Deep learning-based human detection

algorithm was used to detect and recognize human faces (see Fig. 3). It is a powerful open-source recognition project built using Dlib's state-of-the-art facial recognition technique. The algorithm will detect a face in the streaming video sent back by the drone and compare it with faces in the datasets to decide whether it belongs to a stranger.

Fig. 3. Deep learning-based facial recognition

The facial recognition pipeline consisted of the following 4 steps:

(1) Detecting the face. HOG was used to locate faces in an image. To find a face in HOG image, we found the part of the image that looked most like the known HOG pattern extracted from other training faces.
(2) Determining the post of the face. In some cases, a face may turn in different directions and look completely different to a computer. To solve this problem, a face landmark estimation algorithm [11] was used to locate 68 landmarks on every face.

Once we knew where the eyes and mouth were, we simply rotated and scaled the image so that the eyes and mouth were centered as best as possible.

(3) Encoding faces. In order to tell faces apart, a convolutional neural network (CNN) was trained to generate 128 measurements for each face.

(4) Matching the person's name. A simple linear SVM classifier was trained to take measurements from new test images and tell which known person was closest to the match.

3 Experimental Results

Using the computer system (Microsoft Windows 10 PC with a i7-6700HQ CPU, 2.60 GHz per CPU core, 24 GB of memory and a NVIDA GTX 970M video card) with a DJI Mavic Air model drone, the process for one frame with resolution of 640 × 360 pixels took approximately 0.22 s (with resolution of 3840 × 2160 pixels, this took approximately 4.8 s).

3.1 Clear Environment

When the drone kept a horizontal distance from the experimenter, the face could be detected and recognized within 2.95 m. If the drone moved further away, it was unable to recognize the face.

If the drone and the candidate maintained a horizontal distance of 2.95 m, when the candidate turned his/her face to one side (left or right) no more than 45°, the face could be recognized. If the candidate turned his/her face at a larger angle, the face could not be recognized. Also, according to the tests, this angle was independent of the distance between the drone and the face. That is, even if the drone moved in the direction of the face so that the distance between them was closer, the maximum angle did not increase.

At the same distance, if the candidate faced the drone, the maximum height at which the drone recognized his/her face was 1.28 m. If the drone moved to a higher altitude, the face was not recognized. Also, it had an impact on the angle. At this height, the maximum angle at which the drone recognized a human face was 31°. If the angle became larger, the face was not recognized.

3.2 Cloudy

When the drone kept a horizontal distance from the candidate, the maximum distance at which the drone detected and recognized the face was 3 m. If the drone moved further away, it was unable to recognize the face as well.

Since the distance has no particular impact on the angle, we directly tested height in other weather conditions. The maximum height at which the drone recognized a face was 0.8 m. At this height, the maximum angle at which the drone recognized a face was 41°. Similar to sunny days, if the angle became larger, the face was not recognized.

3.3 Sunset

When the drone kept a horizontal distance from the candidate, the maximum distance at which the drone detected and recognized the face was 2.9 m. Also, the maximum height at which the drone recognized a face was 1.71 m. At this height, the maximum angle at which the drone recognized a face was 39°. Similar to other conditions, if the angle became larger, the face was not recognized.

3.4 Dark Conditions with Artificial Light

When the drone kept a horizontal distance from the candidate, the maximum distance at which the drone detected and recognized the face was 2.9 m. Also, the maximum height at which the drone recognized a face was 1.41 m. At this height, the maximum angle at which the drone recognized a human face was 32°. Similar to other conditions, if the angle became larger, the face was not recognized.

3.5 Experimental Summary and Discussions

Table 1 shows the experimental measurements in different conditions: clear environment, cloudy, sunset, and dark under artificial light. Different resolutions were tried during each process. High resolution achieved longer detection distance but resulted in longer processing time. Compared with high resolution, low resolution can achieve faster processing, but it will shorten the detection distance. To achieve real time processing, the resolution was set at 640 * 360.

Table 1. Experimental measurements

	Distance	Height	Angle
Clear environment	2.95 m	1.28 m	31°
Cloudy	3 m	0.8 m	41°
Sunset	2.9 m	1.71 m	39°
Dark conditions with artificial light	2.9 m	1.41 m	32°

4 Conclusions

Today port operations security and property safety have become an important issue of global concern. Many security control systems have been used, such as video surveillance and alarm monitoring, but modern security needs require evolved technology. This research developed a collaborative UAV for surveillance in an unconstrained environment using human facial detection and recognition methods. Based on experimental measurements, the results show that these body detection and facial recognition algorithms perform well under established conditions. Further research will focus on performance improvement, such as extending the ranges of facial recognition within various environments under different conditions.

Acknowledgments. This research was supported by the Center for Advances in Port Management (CAPM) at Lamar University and the Natural Science Foundation (1726500). Any opinions, findings, and conclusions or recommendations expressed in this material are those of the authors and do not necessarily reflect the views of the funding agencies.

References

1. Border Patrol Arrests 123 Men on Smuggling Boats. https://fox5sandiego.com/2019/05/30/border-patrol-arrests-13-men-on-smuggling-boats/
2. Nonami, K., Kendoul, F., Suzuki, S., Wang, W., Nakazawa, D.: Autonomous Flying Robots: Unmanned Aerial Vehicles and Micro Aerial Vehicles. Springer, Tokyo (2010)
3. Eye in the Sky: Abu Dhabi's Ports Now Protected by Drones. https://www.thenational.ae/business/eye-in-the-sky-abu-dhabi-s-ports-now-protected-by-drones-1.595388
4. Dalal, N., Triggs, B.: Histograms of oriented gradients for human detection. In: IEEE Computer Society Conference on Computer Vision and Pattern Recognition, CVPR, San Diego, vol. 1, pp. 886–893 (2005)
5. Viola, P., Jones, M.: Rapid object detection using a boosted cascade of simple features. In: Proceedings of the 2001 IEEE Computer Society Conference on Computer Vision and Pattern Recognition, CVPR, Kauai (2001)
6. Kruppa, H., Santana, M.C., Schiele, B.: Fast and Robust Face Finding via Local Context. IEEE Computer Society, Mantua (2003)
7. Jiang, H., Learned-Miller, E.: Face detection with the faster R-CNN. In: 12th IEEE International Conference on Automatic Face and Gesture Recognition, FG, Washington, D.C., pp. 650–657 (2017)
8. Bootstrapping Face Detection with Hard Negative Examples. https://arxiv.org/pdf/1608.02236.pdf
9. Gaszczak, A., Breckon, T., Han, J.: Real-time people and vehicle detection from UAV imagery. In: Proceeding SPIE, vol. 7878 (2011)
10. Rudol, P., Doherty, P.: Human body detection and geolocalization for UAV search and rescue missions using color and thermal imagery. In: IEEE Aerospace Conference, Big Sky, pp. 1–8 (2008)
11. Kazemi, V., Sullivan, J.: One millisecond face alignment with an ensemble of regression trees. In: IEEE Conference on Computer Vision and Pattern Recognition, CVPR, pp. 1867–1874 (2014)

Novel Techniques of Human-Robot Interaction and Exoskeleton

Humanoid Robotics: A UCD Review

Niccolò Casiddu, Francesco Burlando, Claudia Porfirione, and Annapaola Vacanti(✉)

Dipartimento Architettura e Design, Università di Genova, Genoa, Italy
{niccolo.casiddu,francesco.burlando}@unige.it,
claudia.porfirione@arch.unige.it, annapaola.vacanti@edu.unige.it

Abstract. The research provides a design-driven overview of the fifty main humanoid robots which have been used over the last years in the revolutionary robotics field. The study provides a comparison of the principal aesthetic and interaction features in relation to what kind of environment and users each humanoid robot was designed for.

In order to paint a clearer picture of the humanoid robotics panorama, products are analyzed taking also into account the original purpose they were designed for, with the awareness that humanoid robots often end up being used in sectors they were not specifically meant to.

As a result, the research defines a user-centered taxonomy of humanoid robotics and provides a graphical display of the data about the aesthetic and interaction features of the analyzed robots.

Keywords: User Centered Design · Interaction design · Humanoid robotics

1 Introduction

Due to several issues that society is about to face, such as the increase of the dependency ratio, it is important to design solutions to relieve the burden of such problems from the active population. Unmanned system and robotic devices are helpful tools both for activities dangerous for humans or which humans normally find unpleasant and for the assistance and care of weak users.

With regard to the latter, humanoid robotics have proven to be very efficient, particularly when employed in the care of people with physical or cognitive disorders [1, 2], as in the treatment of physical illness since studies have shown that such robots are able to increase the users' propensity for imitation [3]. On the other hand, a robot designed to operate in industrial workspaces should not be humanlike for any practical purpose. Nevertheless, we are seeing an increasing trend of anthropomorphic design of CoBots. So, it seems clear that at least some humanoids features are always preferred in the design of a machine bond to work at the side of a human.

However, the scope of humanoid robotics is still far from the expectation of the past decades, due to problems both in terms of software and hardware. Nevertheless, these issues didn't stop the diffusion of such products into various areas of our society, especially those involving weak users, as stated before. Therefore, is important for UX

M. Zallio (Ed.): AHFE 2020, AISC 1210, pp. 87–93, 2021.
https://doi.org/10.1007/978-3-030-51758-8_12

and UI designers to increase knowledge about aesthetic and interaction features. Such practice can promote the inclusion of User Centered Design (UCD) and Design for All (DfA) methods in the development of humanoid robots.

2 Humanoid Robotics

Humanoids are robots which are chiefly humanlike in their morphology and behaviour. Eaton [4] outline a detailed taxonomy for n + 1 separate levels of embodied humanoid.

0 Replicant. *Looks exactly like a human being in terms of physical aspect and behavior.*

1 Android. *Very close to human morphology and behavior. Difficult to distinguish from a human.*

n – 3 Humanoid. *Close to a human with high levels of intelligence and dexterity. However, there is no possibility of mistaking the robot for a human being.*

n – 2 Inferior Humanoid. *Has the broad morphology of humans and a reasonable intelligence but may be confined mainly to a limited task set.*

n – 1 Human-Inspired. *Looks quite unlike a human but has the broad morphology of humans. It may have bipedal capabilities or be wheeled. Limited intelligence and dexterity.*

n Built-for-Human. *Looks nothing like a human but it is able to operate in most human-centered, designed environments* [5].

However, such categorization does not take into account several factors, such as body parts. In fact, often robots are very human-like in some parts, as the head, but carry considerable deficits in other parts, as legs are usually absent. Moreover, Eaton taxonomy considers both AI and morphological factors, although if we consider just the latter no robot would ever reach the Replicant status at the present moment. The result is that some levels are empty while some other should be subdivided. Perhaps, looking at the problem on the basis of which users and purpose every robot was designed for could help to rework the taxonomy.

3 Humanoid Robotics Application

Generally speaking, humanoid robots can find application in different contexts, such as:

1. Tasks dangerous for humans
2. Tasks which humans normally find unpleasant or tedious
3. Home assistance
4. Care of weak users
5. Entertainment

Since many robots in scientific fiction look like humans, in the collective imagination robots are humanoids by default. Actually, thanks to advancing technology, recently there has been a trend to design robots increasingly humanlike, triggering the so-called uncanny valley effect [6].

However, with regards to the first two categories described above, a robot designed to operate in industrial workspace should not be humanlike for any practical purpose. Nevertheless, as stated before we are seeing an increasing trend of anthropomorphic design of CoBots. In respect of the home assistance field, apart from virtual assistants like Amazon Alexa or Google Home, humanoid shapes are unavoidable for a humanoid robot aimed at helping people in physical tasks. In fact, in this case both the environment and the tools are human-centered designed. The width of a corridor, the height of a stair, the size and shape of chairs and tables are all designed by referring to ergonomic standards.

Therefore, it should be more economical and reasonable to use humanoid robots whose design is based on these standards than to re-design environments and tools [7]. Moreover, caregiving and entertainment represent areas in which humanoid robotics proves to be more efficient due to the high level of engagement with the user [8].

4 Visualization

This section provides a graphical analysis of the aesthetic and interaction features of fifty humanoid robots. Since humanoid robotics field is constantly growing, only the fifty most interesting products from the last twenty years were selected.

Robots are divided in groups according to the main original purpose they were designed for. *Physical tasks* include all those products which are intended to carry out a bodily work on behalf of a human. In *social assistance* robots are grouped that can be assimilated to a vocal assistant with a body and companion robots. If the robot is specifically designed for the assistance of weak users, you will find it in *caregiving*. *Entertainment* includes all those robots that are intended to act or play sports to entertain people. In *education* you will normally find products designed to teach the foundations of coding to children.

The following features are displayed: year of production, name of the product, country in which the company/research center/university is based, price ($ < 2.000, 2.000 < $$ < 20.000, $$$ > 20.000), conformity to Eaton's taxonomy, degrees of freedom, height, weight, two main colors, presence of legs, arms and eyes, vocal interaction and face recognition (Figs. 1 and 2).

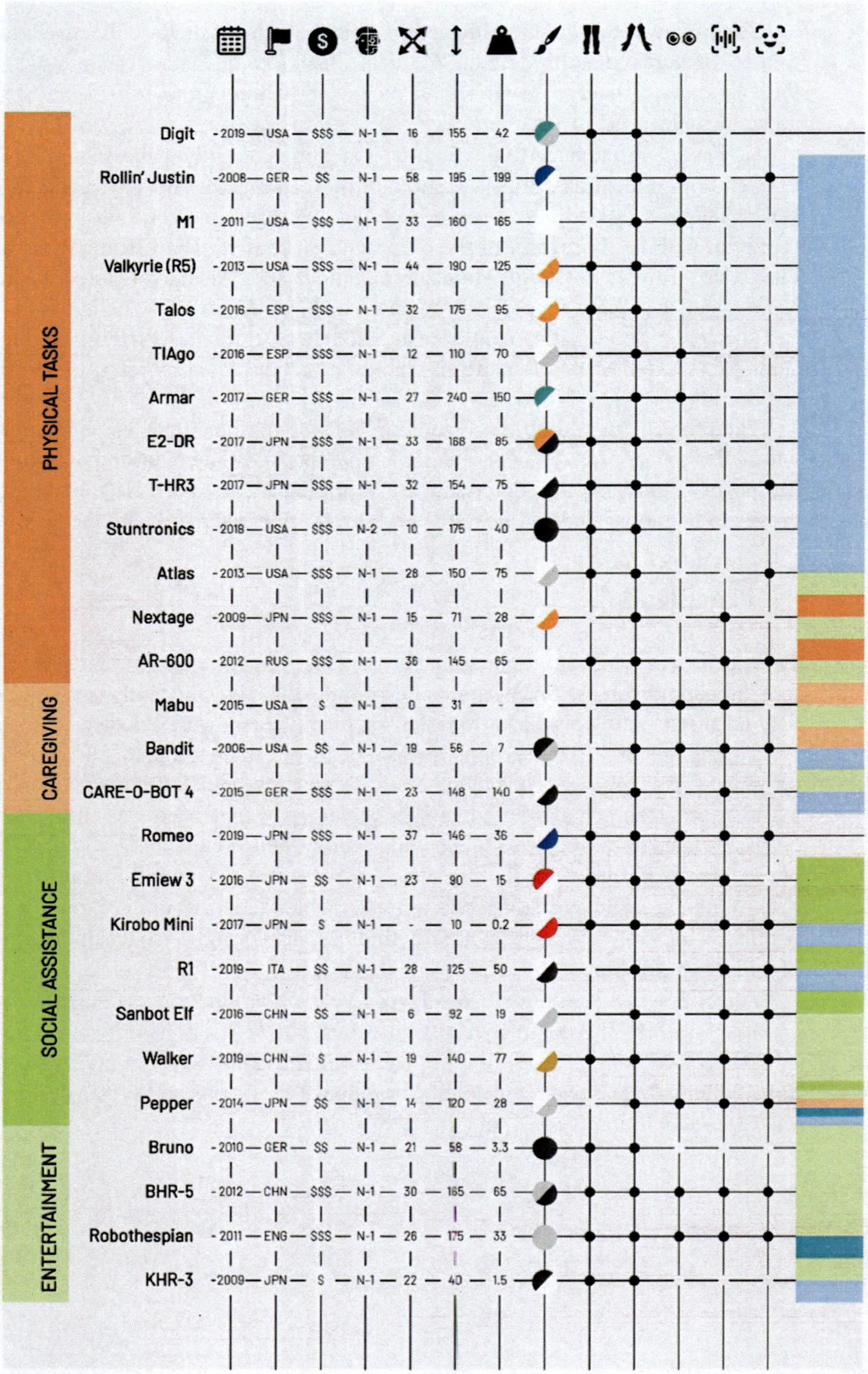

Fig. 1. The charts show fifty robots categorized according to the criteria exposed in the *Visualization* chapter.

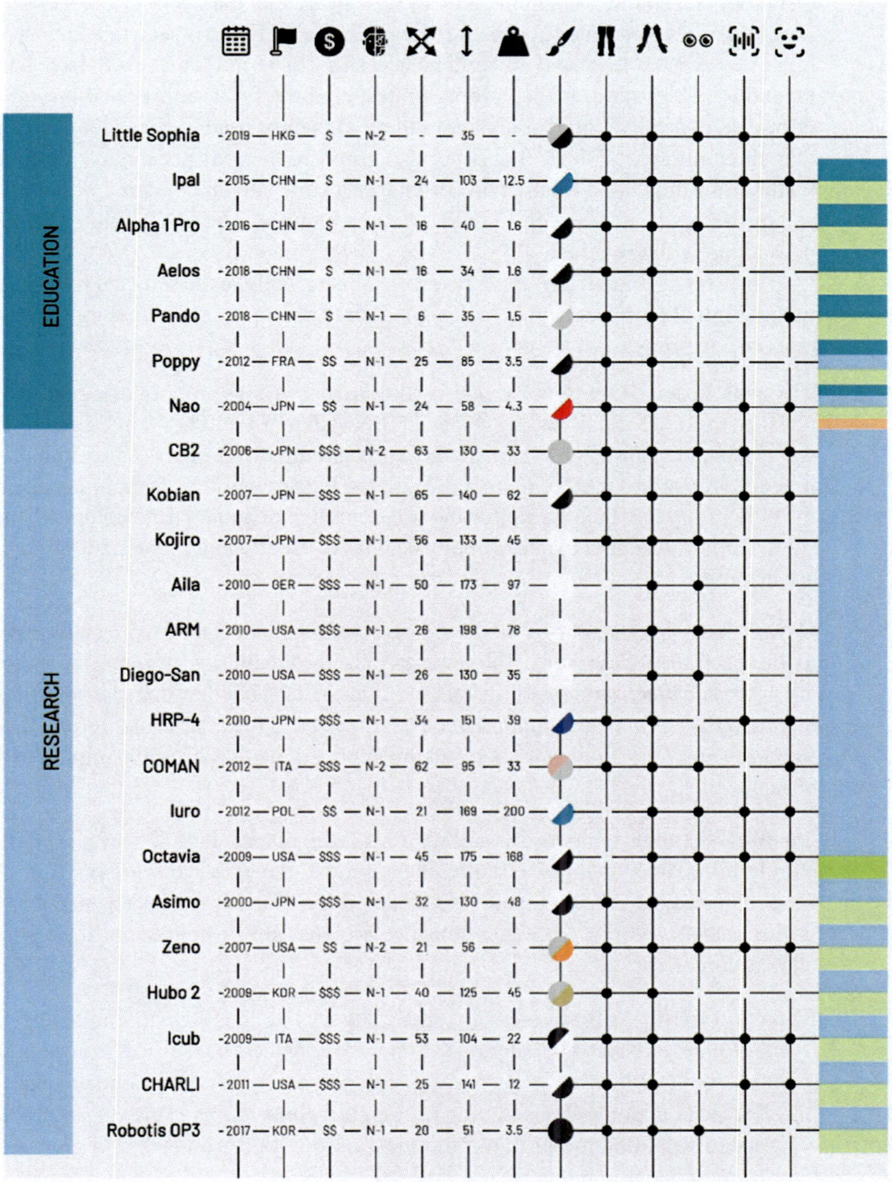

Fig. 2. The charts show fifty robots categorized according to the criteria exposed in the *Visualization* chapter.

5 Findings and Conclusion

The visualization shows that the main purposes for which humanoid robots are designed are research and physical tasks. In most cases such products come at a high price, as a result of the technology behind them and due to the fact that, very often, these robots are prototypes. As mentioned above, the categorization of Eaton is such that almost all robots relate to the $n - 1$ level and it would be necessary to define sublevels to have an accurate taxonomy. Regarding to the colors, white is definitely the most used one. Moreover, even if the visualization displays only the two mainly used, rarely robots present more than three different colors. It appears that arms are the indispensable anthropomorphic feature for a humanoid robot, while often legs are missing in favor of a wheeled structure as physical eyes are frequently replaced by a monitor. The level of interaction usually goes along with the price.

However, as stated above, the research aims to analyze the selected robots according to the original purpose they were designed for. Therefore, it is necessary to examine the outcome as follows.

Physical Tasks. Representing one of the largest groups, robots designed to perform physical tasks are usually of big dimensions and expensive. They sometimes have legs or wheels, but they always have arms with a high number of degrees of freedom. Generally, interaction features are not included. With the exception of *Zeno* from *Research*, this is the only group in which we notice the use of orange or aquamarine colors for the plastic shell. Robots designed to perform physical tasks usually end up being used for research applications.

Caregiving. The category includes robots specifically designed to assist weak user. Due to the specificity of the scope, this appears to be the smallest group. The most used colors are white and black, and robots usually don't have legs, but they do have vocal interaction features. Both *Mabu* and *Bandit* have been truly used for caregiving purposes, although *Research* and *Entertainment* are recurring application areas in conformity with the other groups.

Social Assistance. Robots for social assistance in our sample have been produced mainly in Japan and China. Usually, they are less expensive than robots from *Physical Tasks* albeit vocal interaction and face recognition features are nearly always available. In this category you can notice a lot of robots that are truly used for the purpose they were designed for.

Entertainment. As for *Caregiving*, this category includes a small number of products. Even if *Entertainment* is the biggest group if we consider the real applications, this group includes robots which have been designed specifically for entertainment purposes, as play soccer or act on a stage. As for *Caregiving*, *Entertainment* robots are not colorful, contrary to what one might think: the most used colors are black and grey. All robots have arms and legs.

Education. Typically, this group encloses robots designed to teach the basics of programming to young students. As for *Social Assistance*, those robots are truly used for education purposes, but they are also used for *Entertainment* very often. Usually, they are small and cheap, and they always have arms and legs.

Research. As proof of the fact that we've got a long way to go before humanoid robotics will be a mature technology, *Research* is the largest category if we take into account the original purpose they were designed for and even more considering the real applications. This is also the most varied category, but it can be said that those robots are always prototypes and therefore they are very expensive.

Further research will investigate deeper the subject, taking into consideration a larger number of robots and providing analyses about the upgrade that some of those products showed from one version to the next. Besides, a more specific user-centered taxonomy of humanoid robots' aesthetic and interaction features will be defined.

References

1. Martinez-Martin, E., del Pobil, A.P.: Personal Robot Assistants for Elderly Care: An Overview'. Personal Assistants: Emerging Computational Technologies, edited by Angelo Costa et al., pp. 77–91 Springer International Publishing (2018). https://doi.org/10.1007/978-3-319-62530-0_5
2. Damiani, P.: Robotica educativa e aspetti non verbali nei disturbi specifici di apprendimento. In: Proceedings of 2013 Didamatica Tecnologie e Metodi per la Didattica del Futuro, AICA-Scuola Superiore Sant'Anna-CNR Pisa, pp. 1211–1220. iris.unito.it (2013) https://iris.unito.it/handle/2318/134850#.XG_jvOhKjOg
3. Oberma, L.M., McCleery, J.P., Ramachandran, V.S., Pineda, J.A.: EEG evidence for mirror neuron activity during the observation of human and robot actions toward an analysis of the human qualities of interactive robots. Neurocomputing **70**(13), 2194–2203 (2007). https://doi.org/10.1016/j.neucom.2006.02.024
4. Eaton, M.: Evolutionary Humanoid Robotics. Springer, Berlin (2015). https://doi.org/10.1007/978-3-662-44599-0
5. Brooks, R., et al.: Sensing and manipulating built-for-human environments. Int. J. Humanoid Rob. **1**(01), 1–28 (2004)
6. Mori, M., MacDorman, K.F.: The uncanny valley (from the field). IEEE Rob. Autom. Mag. **19**(2), 98–100 (2012). https://doi.org/10.1109/mra.2012.2192811
7. Kajita, S., et al.: Introduction to Humanoid Robotics. Springer, Berlin (2014). https://www.springer.com/us/book/9783642545351
8. Valentí Soler, M., et al.: Social robots in advanced dementia. Front. Aging Neurosci. **7**, 133 (2015)

Influence of Two Industrial Overhead Exoskeletons on Perceived Strain – A Field Study in the Automotive Industry

Michael Hefferle[1,2](✉), Marc Snell[1], and Karsten Kluth[2]

[1] Department for Occupational Safety and Ergonomics, BMW AG, Moosacher Strasse 51, 80809 Munich, Germany
{michael.hefferle,marc.snell}@bmw.de
[2] Ergonomics Division, University of Siegen, Paul-Bonatz-Straße 9-11, 57068 Siegen, Germany
k.kluth@aws.mb.uni-siegen.de

Abstract. Due to the increasing mean age of workforce across all industry sectors, work-related musculoskeletal diseases, which already have an impact on overall production capacities, are becoming more than ever the focus of attention. Strenuous postures, repetitive tasks, and heavy loads are risk factors for developing work-related diseases. Exoskeletons, which have been suggested as a preventative measure for musculoskeletal disorders, are piloted in various industrial environments. Although psychological and physiological consequences on the wearer have been increasingly investigated, the so far conducted studies mainly focused on the ergonomic evaluation in a laboratory setting. Field studies which evaluate the effects of exoskeletons under real working conditions are scarce. This paper investigates the influence of two different overhead exoskeletons on perceived strain among eight male associates on the assembly line of an automotive manufacturer. Assessment of perceived strain, body part related through Borg's CR-10 scale combined with a modified body map and whole-body through a visual analogue scale (VAS), revealed statistically significant reductions in upper limbs, shoulders (anterior and posterior) as well as neck and spine while using the exoskeletons.

Keywords: Overhead work · Shoulder injury · Work-related musculoskeletal disorders · Exoskeleton · Perceived strain · CR-10 · Visual analogue scale · VAS

1 Introduction

Musculoskeletal disorders (MSD) have been cited as the main reason for sick leaves in the automotive industry [1]. The number of sick days due to MSDs has been calculated to be as high as 125 million for example in Germany [2].

Repetitive tasks that are performed overhead are risk factors for developing diseases of the shoulder [3]. Examples are bursitis, tendinitis or impingement syndrome. MSDs affecting the shoulder account for the longest sick leaves among all work-related musculoskeletal disorders and are therefore particularly problematic [4, 5].

M. Zallio (Ed.): AHFE 2020, AISC 1210, pp. 94–100, 2021.
https://doi.org/10.1007/978-3-030-51758-8_13

Lately, Exoskeletons have been seen increasingly attention as a preventive method in various industry sectors, as they are intended to reduce strains on specific body parts by redirecting external loads to less strained body regions or by retaining favorable ergonomic postures [6]. Although having been introduced in various production environments, only a few studies so far have evaluated the physiological impact on the wearer [7]. The majority of the studies were set up in a laboratory environment and focused primarily on assessing muscle activation on a few distinct muscle groups and/or motion analysis and/or perceived effort [8–11]. There is enough evidence to support the hypothesis that psychological and physiological consequences on the wearer might change in a real work environment. Hence evaluations of exoskeletons should not be performed exclusively in a laboratory setting [12].

This paper evaluates two different passive industrial exoskeletons to support overhead work in the assembly department of an automotive manufacturer, by assessing perceived body part and whole-body strain.

2 Methods

2.1 Exoskeleton Devices

In this study two commercially available upper limb exoskeleton devices, the Crimson Dynamics (Model 2019, Crimson Dynamics Technology Co. Ltd., Dalian, China) and the Skelex V1 (Model 2017, Skelex, Rotterdam, The Netherlands), as shown in Fig. 1 were investigated.

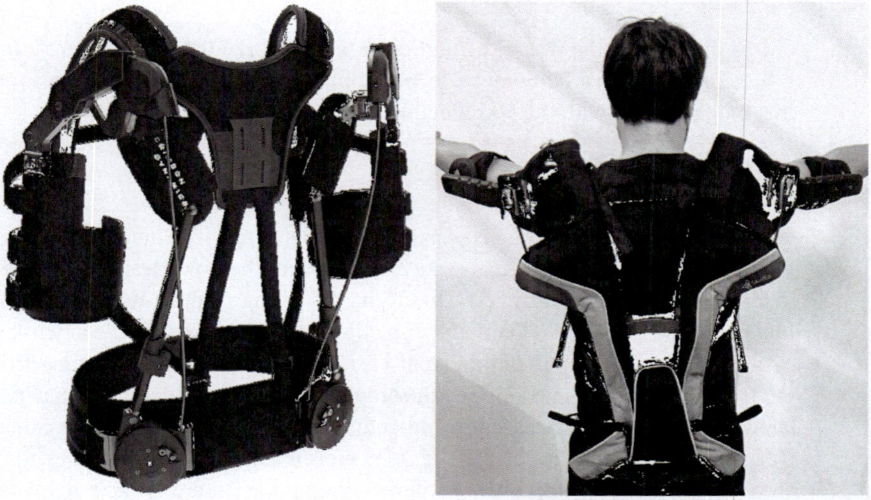

(a) *Crimson Dynamics (Exo1)* (b) *Skelex V1 (Exo2)*

Fig. 1. (a) Crimson Dynamics (www.c-dyn.com) and (b) Skelex V1 (www.venturesone.com).

Upper limb exoskeleton devices aim to almost compensate the weight of the arms. They are intended to reduce the strain exerted on the shoulders while working in an overhead posture. The term overhead work is typically used to describe tasks where the arms are raised at or even above shoulder level [13]. Upper limb exoskeletons are usually worn like a backpack, while the upper arms rest in two cuffs that are each fixed to a lever. The two levers run parallel to the upper arm and are connected to the main frame through joints next to the shoulder, which serve as pivot points for the mechanism. The main frame, often a rigid structure running parallel to the wearer's spine, redirects the loads and strains experienced by the upper limbs directly to the hips, thereby bypassing the shoulders' muscles, ligaments, and joints that are most likely to experience injuries.

Passive exoskeleton devices rely mainly on spring mechanisms or other types of energy storage to release the necessary force to compensate for the weight of the wearer's arms.

2.2 Participants

A convenience sample of eight male workers were recruited from the automotive assembly department for the field study. Their anthropometric data (mean ± standard deviation) are included in Table 1. All recruited participants had no history of shoulder injury of any kind. Written consent was obtained.

Table 1. Anthropometric data of participants.

Participants	Age [y]	Statue [cm]	Weight [kg]	BMI [kg/m^2]
8 m	37.5 ± 13.0	183.1 ± 3.4	94.0 ± 8.6	28.1 ± 3.4

At the time of the study, the participants had worked at their current job a mean of 5.8 ± 3.7 years. On a 10-point scale, participants rated their physical fitness at 5.6 ± 1.6 in a range of 3–8, with the 10 reflecting the highest level of fitness.

2.3 Description of Workplace

The investigation took place in the assembly department of an automobile manufacturer, where overhead work is typically performed. After the powertrain and body have been merged the exhaust is installed. Due to media filling (lubricants, operating fluids and fuel) the body of the car can no longer be tilted, which makes overhead work inevitable. While the heavier rear part of the exhaust is raised to the underbody using a lifting device, the front part is lifted manually by the employee and attached to the exhaust pipe of the engine. Afterwards, light assembly procedures are carried out before the exhaust system is screwed to the underbody. The tasks performed at the workstation result in a mean shoulder joint angle of roughly 90° depending on the stature of the employee.

2.4 Experimental Procedure

For each experimental condition, and immediately following the fitting of the exoskeleton in the case of conditions *Exo1* and *Exo2*, participants began their work cycles as usual.

Data were collected until the participant had performed at least ten complete work cycles (i.e. without any interruptions) for each experimental condition. This criterion was not met for three conditions out of total 24 (3 conditions × 8 participants), due to unforeseen disturbances such as conveyer downtime. Every participant completed subjective evaluations after each experimental condition by filling in a paper-based questionnaire. Following a within-subject design, each participant went through the same test design three times, once without (*NoExo*) and twice with intervention (*Exo1*, *Exo2*). The order of the experimental conditions was randomized across participants, to avoid order and learning effects.

2.5 Data Collection and Analysis

Perceived local strain was assessed by using Borg's CR-10 scale [14] applied on a modified body map [15]. Perceived whole-body strain was evaluated using a Visual Analogue Scale (VAS) [16] represented as a line with verbal and visual anchors, "easy" and "exhausting", and corresponding smiley faces [17] at either end. Relative changes in perceived strain for each body part and the whole body were calculated using formula (1).

$$Relative\ change_i = \frac{Mean_{i,intervention} - Mean_{i,control}}{10} * 100\%. \quad (1)$$

Statistical analyses were performed in Minitab® (Version 18.1, Minitab, Inc., Pennsylvania 16801, USA). Non-parametric Wilcoxon-Signed-Rank-Tests were performed between *NoExo – Exo1* and *NoExo – Exo2* conditions, with a Bonferroni-corrected significance level set at $\alpha = .025$.

3 Results

Results of the subjective evaluation revealed a statistically significant relative reduction of the right elbow (−14%, $p = .024$) between *NoExo* and *Exo2* condition. Mean values for *shoulders, anterior* were statistically significant reduced by −23% (right) and −20% (left) between *NoExo* and *Exo1* condition and respectfully showed a trend, although not statistically significant, toward reduction between the control and intervention *Exo2*. The remaining body parts for the upper limb region *hand & fingers, forearm, elbow* (only left side) and *upper arm* showed trends of a relative reduction (between −8% and −14%) in perceived strain between the two intervention and control conditions. However, none of these trends were statistically significant. The same applied for the body parts *upper torso* and *lower torso*, which showed relative reduction rates between −1% and −6%. For *shoulders, anterior*, a statistically significant reduction was found for the right side (−21%, $p = .024$) in intervention condition *Exo1*. *Exo2* showed a relative reduction of −18% for the right side. For the left side, both interventions reduced perceived strain by −21% (*Exo1*) and −16% (*Exo2*). A statistically significant reduction (−16%, $p = .017$) was found for the *neck* in intervention condition *Exo2*. Perceived strain was reduced by −18% for condition *Exo1*. The *shoulders, posterior* showed statistically significant strain reductions for both left (−21%, $p = .020$) and right (−23%, $p = .020$) side in intervention condition *Exo1*. In condition *Exo2* tendencies of reduction for the right (−19%) and left side (−16%) were found. The *spine* showed statistically significant reductions for both intervention conditions (*Exo1*: −19%, p = .017; *Exo2*: −20%, p = .016) exclusively. Results are shown as box plots in Fig. 2.

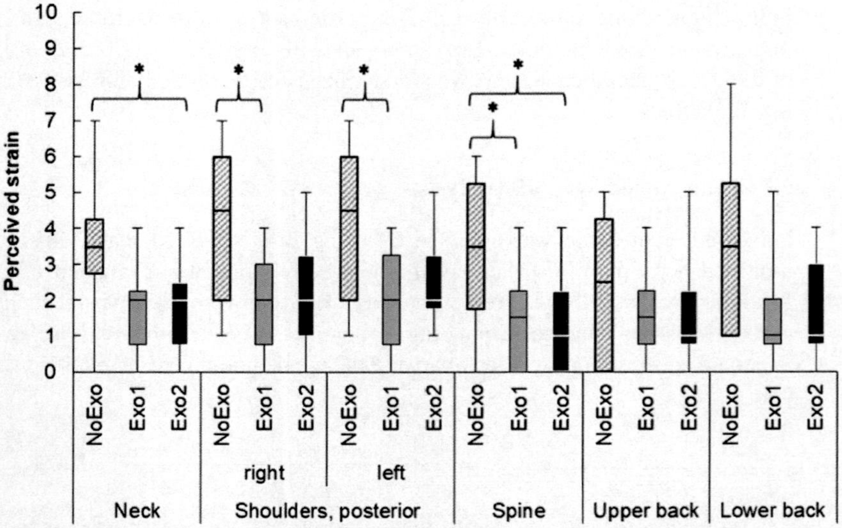

Fig. 2. Perceived strain for *neck, shoulder, posterior, spine, and back*. "0": no strain, "10": very high strain (*: p < .025).

Upper back (*Exo1*: −8%, *Exo2*: −8%) and *lower back* (*Exo1*: −21%, *Exo2*: −20%) showed tendencies of reduction, although not being statistically significant. Perceived whole-body strain decreased statistically significant for the condition *NoExo - Exo1* (−38%, p = .017). For the condition *NoExo - Exo2* by −22%.

4 Discussion

The results demonstrated statistically significant reductions in the body parts supported by the exoskeleton, confirming previous results using a similar device in an automotive industrial setting [18]. However, contrary results have been demonstrated in another field study in the automotive industry which investigated the influence of an overhead exoskeleton on perceived strain in a real working situation. Here, significant strain reductions on body regions not directly supported by means of the overhead exoskeleton device (from a technical-functional point of view the *elbows*, *forearms*, *hand & fingers* are not directly supported). Interestingly, that investigation found no differences in perceived strain for the "shoulders" or the "back" [19]. The results presented here do not align with the aforementioned study. Superficially similar, both studies were carried out in the field and the same industrial sector. The investigated workplaces differed slightly though (different cycle times and different tasks) and a different exoskeleton device was used. However, the present results regarding perceived strain are confirmed by further studies [20, 21] and therefore appear to be valid. The aforementioned studies reveal differences in reported perceived strain across studies although being conducted in similar experimental settings. Therefore, further investigations should be performed in field and laboratory environments – ideally having similar or even replicated study designs – to address the problem of small sample sizes which may have led to the statistical tests having a low power [21], thereby resolving the issue of conflicting results.

5 Conclusion

This paper reviews previously conducted studies on industrial exoskeleton devices which support overhead work and are commercially available. Findings were that most of the published studies have been conducted in laboratory environments. Results of perceived body part and whole-body strain, acquired through Borg's CR-10 scale combined with a modified body map as well as a Visual Analogue Scale, revealed significant reductions for *shoulders, anterior* (right), *shoulders, posterior*, *spine* and whole-body in the case of the Crimson Dynamics's device (*Exo1*). Significant reductions for the *elbow* (right), *neck* and *spine* were found for the Skelex exoskeleton device (*Exo2*).

References

1. Statistical office of the European communities: Health and safety at work in Europe (1999–2007). A statistical portrait. Eurostat. Statistical books. Office for official publications of the European Union, Luxembourg (2010)
2. Bundesanstalt für arbeitsschutz und arbeitsmedizin berufskrankheiten durch mechanische einwirkungen (2017)
3. Bjelle, A., Hagberg, M., Michaelsson, G.: Clinical and ergonomic factors in prolonged shoulder pain among industrial workers. Scand. J. Work Environ. Health **5**(3), 205–210 (1979). https://doi.org/10.5271/sjweh.3094
4. American Society of Biomechanics (Hrsg): EMG assessment of a shoulder support Exoskeleton during on-site job tasks (2017)
5. Bargende, M., Reuss, H.-C., Wiedemann, J. (Hrsg): 17. Internationales Stuttgarter Symposium. Automobil- und Motorentechnik. Proceedings. Springer Fachmedien Wiesbaden, Wiesbaden (2017)
6. de Looze, M.P., Bosch, T., Krause, F., Stadler, K.S., O'Sullivan, L.W.: Exoskeletons for industrial application and their potential effects on physical work load. Ergonomics **59**(5), 671–681 (2016). https://doi.org/10.1080/00140139.2015.1081988
7. Weston, E.B., Alizadeh, M., Knapik, G.G., Wang, X., Marras, W.S.: Biomechanical evaluation of exoskeleton use on loading of the lumbar spine. Appl. Ergon. **68**, 101–108 (2018). https://doi.org/10.1016/j.apergo.2017.11.006
8. Huysamen, K., Bosch, T., de Looze, M., Stadler, K.S., Graf, E., O'Sullivan, L.W.: Evaluation of a passive exoskeleton for static upper limb activities. Appl. Ergon. **70**, 148–155 (2018). https://doi.org/10.1016/j.apergo.2018.02.009
9. Kim, S., Nussbaum, M.A., Mokhlespour Esfahani, M.I., Alemi, M.M., Alabdulkarim, S., Rashedi, E.: Assessing the influence of a passive, upper extremity exoskeletal vest for tasks requiring arm elevation. Part I - "Expected" effects on discomfort, shoulder muscle activity, and work task performance. Appl. Ergon. (2018). https://doi.org/10.1016/j.apergo.2018.02.025

10. Kim, S., Nussbaum, M.A., Mokhlespour Esfahani, M.I., Alemi, M.M., Jia, B., Rashedi, E.: Assessing the influence of a passive, upper extremity exoskeletal vest for tasks requiring arm elevation. Part II - "Unexpected" effects on shoulder motion, balance, and spine loading. Appl. Ergon. (2018). https://doi.org/10.1016/j.apergo.2018.02.024
11. Muramatsu, Y., Kobayashi, H., Sato, Y., Jiaou, H., Hashimoto, T., Kobayashi, H.: Quantitative performance analysis of exoskeleton augmenting devices - muscle suit - for manual worker. Int. J. Autom. Technol. **5**(4), 559–567 (2011). https://doi.org/10.20965/ijat.2011.p0559
12. Ferguson, S.A., Allread, W.G., Le, P., Rose, J., Marras, W.S.: Shoulder muscle fatigue during repetitive tasks as measured by electromyography and near-infrared spectroscopy. Hum. Factors **55**(6), 1077–1087 (2013). https://doi.org/10.1177/0018720813482328
13. Bier, M.: Ergonomie der Überkopfarbeit. Zugl.: Darmstadt, Techn. Hochsch., Diss. Fortschritt-Berichte VDI @Reihe 17, Bd 70. VDI Verl., Düsseldorf (1991)
14. Borg, G.: Borg's Perceived Exertion and Pain Scales. Human Kinetics, Champaign (1998)
15. Corlett, E.N., Bishop, R.P.: A technique for assessing postural discomfort. Ergonomics **19**(2), 175–182 (1976). https://doi.org/10.1080/00140137608931530
16. Kersten, P., Küçükdeveci, A.A., Tennant, A.: The use of the visual analogue scale (VAS) in rehabilitation outcomes. J. Rehabil. Med. **44**(7), 609–610 (2012). https://doi.org/10.2340/16501977-0999
17. Kluth, K.: Analyse, Beurteilung und ergonomische Gestaltung von Arbeitsplätzen in Selbstbedienungsläden. Höpner und Göttert, Siegen (2001)
18. Spada, S., Ghibaudo, L., Gilotta, S., Gastaldi, L., Cavatorta, M.P.: Analysis of exoskeleton introduction in industrial reality: main issues and EAWS risk assessment. In: Proceedings of the AHFE 2017 International Conference on Physical Ergonomics and Human Factors. Springer, Heidelberg (2018)
19. Hefferle, M., Dahmen, C., Kluth, K.: Einfluss eines Exoskeletts zur Unterstützung von Überkopftätigkeiten in der Automobilindustrie auf die subjektive, körperliche Beanspruchung. Eine explorative Feldstudie. ASU - Arbeitsmedizin, Sozialmedizin, Umweltmedizin (12) (2019)
20. Hensel, R., Keil, M.: Subjektive Evaluation industrieller Exoskelette im Rahmen von Feldstudien an ausgewählten Arbeitsplätzen. Z. Arb. Wiss. **72**(4), 252–263 (2018). https://doi.org/10.1007/s41449-018-0122-y
21. Kim, S., Nussbaum, M.A.: A follow-up study of the effects of an arm support exoskeleton on physical demands and task performance during simulated overhead work. IISE Trans. Occup. Ergon. Hum. Factors 1–12 (2019). https://doi.org/10.1080/24725838.2018.1551255

Human-Robot Interaction via a Virtual Twin and OPC UA

Christoph Zieringer, Benedict Bauer, Nicolaj C. Stache, and Carsten Wittenberg(✉)

Department Mechanics and Electronics, Heilbronn University of Applied Sciences, 74081 Heilbronn, Germany
{benedict.bauer,nicolaj.stache, carsten.wittenberg}@hs-heilbronn.de

Abstract. The idea of a digital factory is closely combined with the concept of virtual or digital twins. The development and implementation of virtual twins is facilitated by the so-called game engines which offer huge possibilities in the virtual space. These game engines provide programming interfaces to several development environments. This paper describes early steps of the development of a virtual twin for a 6-axle industrial robot including a user interface. This user interface was evaluated. The results of the evaluation are the basis for a bold redesign, which is described in this paper shortly.

Keywords: Virtual twin · Virtual engineering · Virtual reality · Human-robot interaction · Robotics · IIoT

1 Introduction

Developing an industrial production line is the entering wedge for the fabrication of a new product. The lifecycle of such a production line (Fig. 1) - and thus the engineering - is embedded in long ranged corporate strategy as well as in tactical product development and product planning and control processes [1, 2]. This paper has the scope on the development phase (engineering), using the example of an industrial robot in a laboratory application.

Fig. 1. Lifecycle of an industrial production line [3, 4].

Generally, the development of an industrial and highly-automated production line is a kind of system modelling including the system structure and all the different operating modes. It includes the specification of necessary hard- and software, and also

M. Zallio (Ed.): AHFE 2020, AISC 1210, pp. 101–107, 2021.
https://doi.org/10.1007/978-3-030-51758-8_14

the engineering of the spatial structure. During the basic engineering a comprehensive concept across all trades is outlined. This concept is the frame for the following detail development by each participating engineer.

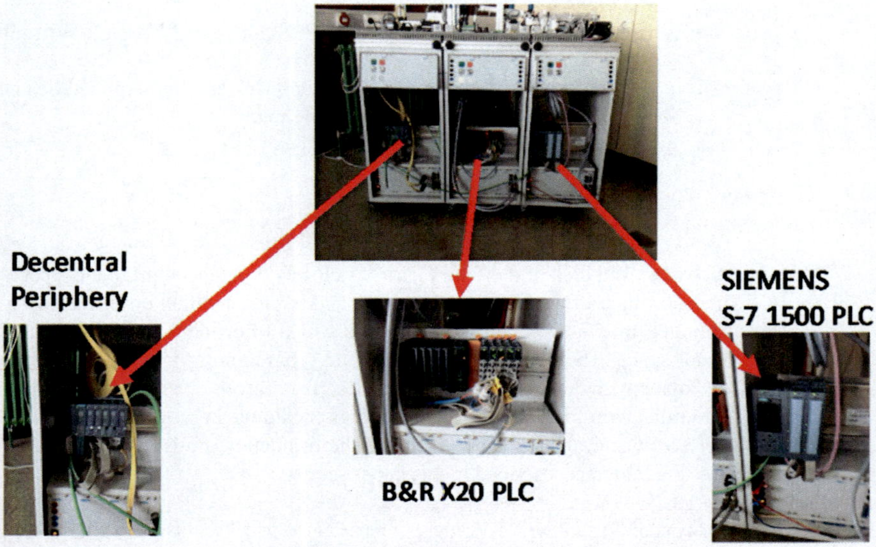

Fig. 2. Sorting system with its PLC

A huge part of the engineering is the software programming containing the development, implementation and testing for all automation devices (e.g. PLCs, Robots, IPCs, etc.). Typically, these tasks are done by the engineers at their desks, except the real commissioning. By now, taking the production line into operation will be done on-site - typically under deadline pressure. At that time a huge number of errors are detected and have to be fixed in a very short time – until the handover to the customer.

2 Earlier Development of a Virtual Twin Application

During the last years the augmented and virtual reality devices made a leap forward in development. Based on this developmental state the usage of these augmented or the virtual reality devices could support the users in the development or engineering phase. Consequently, the concept of a virtual or a digital twin is useful [5].

As a first development of a virtual twin of a simple sorting system in the automation laboratory of Heilbronn University was designed (Fig. 2), realized and tested. This educational sorting system is used for teaching automation engineering with undergraduate students and can distribute circular parts by its colors. It is guided by two programmable logic controllers from different manufacturers (Siemens and B&R). For communication OPC/UA [6] I used.

The virtual engineering lab is located close to the automation laboratory with the sorting system so that network-based time delay issues should not have a huge influence.

As a virtual reality device, an HTC Vive device is used and Ethernet-based connected via a cloud server to the sorting system in the automation laboratory (Fig. 3).

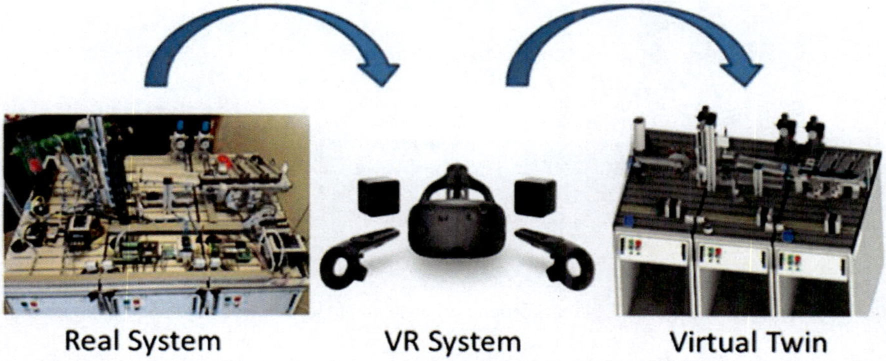

Fig. 3. Mechatronic system in the automation laboratory: real system and virtual twin

The virtual environment of the sorting system is presented to the user by the wired head-mounted display (HMD) of the HTC Vive. Any influences from the outside like sunlight, reflections or any contrast issues do not play a role while using this HMD.

The project outcome of this application showed the possibilities of the use of a virtual twin - the combination of real systems and virtual applications. The results encouraged a transfer of the topic of the use of virtual reality issues to a more complex application which includes also an enhanced movement part. Such a domain could be the programming of industrial robots.

3 IIoT/Industry 4.0 Laboratory Factory at Heilbronn University

Heilbronn University owns and uses a smart production system (Figs. 4 and 5) in the Otto Rettenmaier-Research-Laboratory for digitalization, research and educational purposes.

Liquids and solids are filled automatically in small bottles. These bottles are collected in trays and carried to a recycling station. In this station a robot with six axles empties these bottles and feeds the empty bottles again into the process. This smart production system is described in detail in [7].

The robot of the recycling station is a six-axle buckling arm robot. This robot is used as an example for the development of a digital twin. The modelling of the virtual twin is performed with the Unreal Engine 4. Based on a CAD-Drawing of the robot the drawings of the robot components were converted and imported to the development environment of the Unreal Engine. Enriched with realistic motion values like axles the robot model can move in the virtual space similar to the original robot in the recycling station [8].

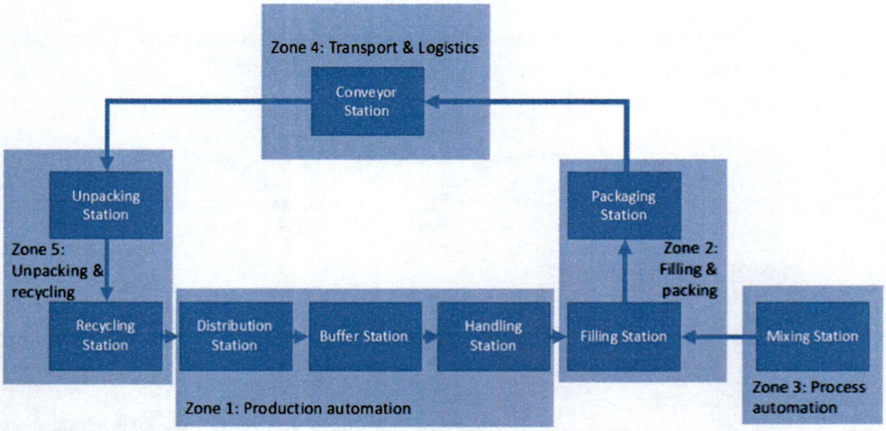

Fig. 4. Structure of the smart production system at Heilbronn University of Applied Science [7]

Fig. 5. Smart production system with the robot

4 Development and Implementation of the Virtual Twin

At the beginning, useful use-cases were identified. Different use cases for virtual twins are feasible. A virtual twin can be used in the engineering phase for the development and implementation of systems without the real hardware, e.g. if the hardware is built in parallel. Another usage can be the simulation of the real system e.g. for predictive maintenance. Tele-operation is also another feasible usage: The user performs operation with the virtual robot in the virtual world. Simultaneously the real robot performs the similar operations in the real world. As a first usage case this research project focuses on the tele-operational scenario.

4.1 Communication Concept

As a communication standard, OPC UA was chosen. This decision assumed that OPC UA will be the future communication standard in the field of IIoT (Industrial Internet of Things) /Industry 4.0. Another reason is that OPC UA is standardly implemented on most PLCs.

The virtual twin is realized as an OPC UA client. The OPC UA server is on a PLC, the data exchange between the virtual twin and the PLC is done by the PLC variables (Figs. 6 and 7).

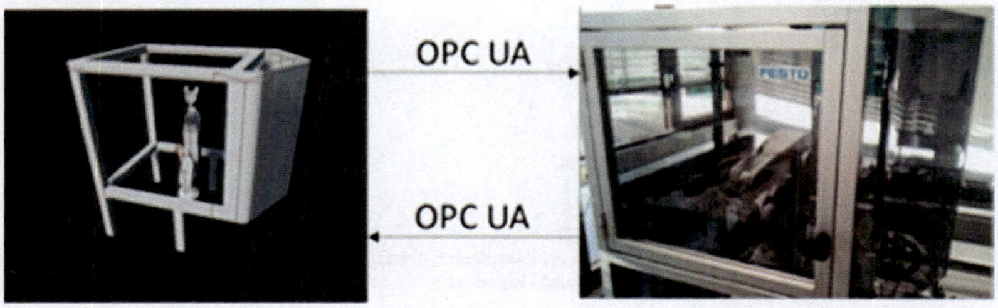

Fig. 6. Simplified communication model between virtual twin and real robot

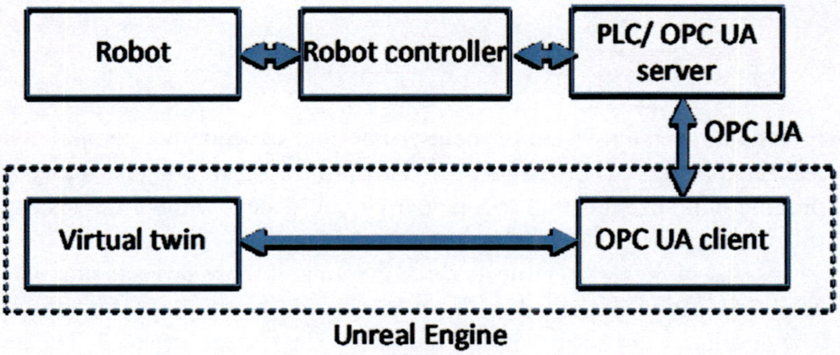

Fig. 7. Communication architecture

In this project, the communication between the virtual and real world is essential because the calculation of the movement of the robot axles is done by the real robot controller. That means that the user input (on the PC with the virtual twin) is sent via OPC UA and the PLC to the robot controller (Fig. 7).

After the calculation of the robot reaction the values are sent back via the PLC and OPC UA to the virtual twin and the virtual twin performs the movement. This is necessary because the development of a "virtual" robot controller is not feasible in a short project duration and is not in the project focus.

4.2 Interface Concept

The first implemented user interface concept has a key-based interaction. The numeric keys are used for the axle selection (key 1 for axle 1 et cetera). The keys L and R are used for the rotating movement of the selected axle.

Further keys are used for the navigation in the virtual room. And for setting movement parameters like speed and acceleration. This UI concept had its only purpose in testing the system functionality. As a consequence, a user-friendly user interface concept was necessary. Based on an evaluation, an improved user interface was developed (Fig. 8).

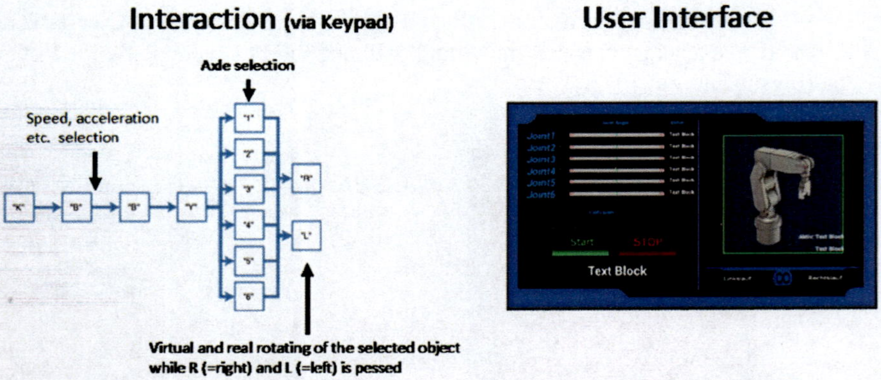

Fig. 8. Interaction and User Interface

5 Outlook

In addition to the revision of the user interface concept in a parallel concept a virtual tech-in concept was developed [9]. This project shows the possibility of the teach-in programming of industrial robots in a virtual space – without the necessity of the real robotic hardware.

A goal of these two projects is the merging of those projects into one environment. The virtual twin described in this paper should be the basis for the virtual teach-in. The development platform is still the same: The Unreal Engine 4. The user interface is different: The virtual twin is developed on an usual but powerful PC; the virtual teach-in uses an HTC Vive device.

References

1. Felix, H.: Unternehmens- und Fabrikplanung: Planungsprozesse, Leistungen und Beziehungen. REFA Fachbuchreihe, Hanser Verlag, Munich (1998)
2. During, A., Komischke, T., Wittenberg, C., Berger, U.: A vision for an information management tool for plant engineering – functionality and user interface. In: Horváth, X. (eds) Proceedings of the TMCE 2004, pp. 12–16. Rotterdam Millpress, Lausanne (2004)

3. Wittenberg, C.: Cause the trend Industry 4.0 in the automated industry to new requirements on user interfaces? In: Kurosu, M. (ed.) Proceedings of Human-Computer Interaction - Users and Contexts, pp 238–245. Springer, Heidelberg (2015)
4. Wittenberg, C.: Human-CPS interaction – requirements and mobile human-machine interaction methods for the Industry 4.0. In: Proceedings of the 13th IFAC/IFIP/IFORS/IEA Symposium on Analysis, Design, and Evaluation of Human-Machine Systems, Kyoto, 30 August–3 September, IFAC-PapersOnLine, vol. 49, issue 19, pp. 420–425 (2016)
5. Armendia, M., Alzaga, A., Peysson, F., Fuertjes, T., Cugnon, F., Ozturk, E., Flum, D.: Machine tool: from the digital twin to the cyber-physical systems. In: Armendia, M., Ghassempouri, M., Ozturk, E., Peysson, F. (eds.) Twin-Control, pp. 3–21. Springer, Berlin (2019)
6. Lange, J., Iwanitz, F., Burke, T.: OPC – From Data Access to Unified Architecture. VDE VERLAG GMBH, Berlin (2010)
7. Wittenberg, C., Bauer, B., Stache, N.: A smart factory in a laboratory size for developing and testing innovative human-machine interaction concepts. In: Ahram, T., Falcão, C. (eds.) Advances in Usability and User Experience, AHFE 2019. Advances in Intelligent Systems and Computing, vol 972, pp. 160–166. Springer, Cham (2020)
8. Zieringer, C., Bauer, B., Wittenberg, C.: Erstellung eines virtuellen Zwillings mit OPCUA und der Unreal Engine 4 4 (Implementation of a virtual twin with OPC UA and the unreal engine). In: Bauer, B., Wittenberg, C. (eds.) Tagungsband AALE. VDE-Verlag, Berlin (2019)
9. Ehmann, D., Wittenberg, C.: The idea of virtual teach-in in the field of industrial robotics. In: Proceedings of 2018 IEEE 14th International Conference on Control and Automation (ICCA), Anchorage, USA, pp. 680–685 (2018)

Self-assessment of Proficiency of Intelligent Systems: Challenges and Opportunities

Alvika Gautam(✉), Jacob W. Crandall, and Michael A. Goodrich

Computer Science Department, Brigham Young University, Provo, UT 84604, USA
alvikag@byu.edu, {crandall,mike}@cs.byu.edu

Abstract. Autonomous systems, although capable of performing complicated tasks much faster than humans, are brittle due to uncertainties encountered in most real-time applications. People supervising these systems often rely on information relayed by the system to make any decisions, which places a burden on the system to self-assess its proficiency and communicate the relevant information.

Proficiency self-assessment benefits from an understanding of how well the models and decision mechanisms used by robot align with the world and a problem holder's goals. This paper makes three contributions: (1) Identifying the importance of goal, system, and environment for proficiency assessment; (2) Completing the phrase "proficient ‹preposition›" using an understanding of proficiency span; and (3) Proposing the proficiency dependency graph to represent causal relationships that contribute to failures, which highlights how one can reason about their own proficiency given alterations in goal, system, and environment.

Keywords: Proficiency · Self-assessment · Goal(s) · System · Environment · Intelligent agents

1 What is Proficiency Assessment?

Proficiency assessment can be operationally defined as *the ability to detect or predict success (or failure) towards a goal in a particular environment given an agent's sensors, computational reasoning resources, and effectors.* Ideally, proficiency assessment approaches need to work a priori, in situ, and a posteriori. Different levels of self-assessment include (a) detecting proficiency (success or failure), (b) assigning a proficiency score (quantification of likelihood of success or degree of failure), (c) providing explanations (reasoning behind the outcome i.e., success or failure), and (d) predicting proficiency, which will allow intelligent systems to make informed decisions about their ability to accomplish tasks based on previous outcomes and their explanations. Communities in both computer science and robotics have addressed questions related to proficiency self-assessment. These include introspection [1–3], monitoring system performance [4, 5], robustness to uncertainties [6, 7], to name a few. However, much work is needed to address more in-depth levels of proficiency self-assessment adequately.

This paper presents initial ideas and frameworks for reasoning about proficiency self-assessment. Proficiency must be defined relative to the following: (i) Goals: desired

M. Zallio (Ed.): AHFE 2020, AISC 1210, pp. 108–113, 2021.
https://doi.org/10.1007/978-3-030-51758-8_15

outcomes of the task that the intelligent agent should reach in a finite amount of time or during long-duration missions; (ii) Environments: world settings in which the intelligent agent operates; and (iii) System Configurations: sensors, actuators, and computational resources that are available to the agent. This paper identifies a causal relationship between these three categories and the mechanisms and models used by AI algorithms to make decisions. The paper is limited to *in situ* aspects (during runtime) of simple detection of proficiency, i.e., whether an agent's actions and the resulting states take it closer towards the desired outcome of the task.

2 Span of Proficiency Self-assessment

Because proficiency depends on the environment, system, and goal, we propose that proficiency assessments should explicitly assert the span for which proficiency applies. We propose that a way to think about span is to consider the following statement:

An agent is proficient ‹*preposition*›, where the ‹*preposition*›, is used to indicate the span or scope of the assessment.

Table 1 proposes a relationship between proficiency span and a representative preposition. The entries of the table represent the instances of variation of these properties (Goal, System, and Environment) over an enumerated set. A “1” in the table indicates that proficiency is defined with respect to a specific goal, specific system, or specific environment, whereas “> 1” indicates multiple goals, systems, or environments.

Table 1. Span of proficiency self-assessment

	Goal	System	Environment
At	1	1	1
Within	1	>1	>1
Across	>1	1	1
Over	>1	>1	>1

We recognize that the selection of the prepositions is somewhat arbitrary, but we believe the prepositions provide a common vocabulary for proficiency researchers – we have been part of too many conversations where people were talking past each other because of confusion about the way they were using the term *proficient.*

Proficient At: An agent is said to be “proficient at” a goal if it competently satisfies the goal for a given system configuration and environment condition. Proficiency “at” this level is the minimum requirement for an agent. The “1”s in each cell of this row in Table 1 highlights that being proficient “at” something appertains only to a single environment, system, and environment.

Proficient Within: Most systems are subject to uncertain and dynamic environmental conditions or disturbances during the task. This is aggravated by uncertainties associated with the system itself, e.g., noisy sensors, failing effectors, time bounds on computation processes. For a given goal, an agent is said to be "proficient within" a range of system configurations and environment conditions. This is represented by the ">1" entries in the "Within" row of the table.

For example, consider a physical robot system that can knowingly fail in three different ways. Hence, for the agent to adequately assess its proficiency towards a goal in the presence of system anomalies, 3! system configurations should be tested. Similarly, an enumerated set of expected variations or changes to the environment is represented by the ">1" entries in the Environment column.

Proficient Across: A system might be comprised of several subsystems working in tandem, but each for their own individual goals. Or a single system might be capable of pursuing multiple goals. "Proficient across" indicates that proficiency needs to be considered with respect to multiple goals, which is represented by the ">1" entry in the "Across" row of the table.

Proficient Over: The "Over" row of Table 1 summarizes the span across which assessment of proficiency should ideally hold. That is, an intelligent agent should be able to assess its ability to competently satisfy multiple goals over a range system configurations, worlds and environment settings.

3 Proficiency Dependency Graph

In our approach towards self-assessment of proficiency, we adopt a proficiency dependency graph (PDG). The graph has six vertices: V = (Outcome, Mechanism, Model, World/Environment, System/Physical robot, Goal) as shown in Fig. 1. A directed edge connects vertex *A* with vertex *B* if " B depends on A."

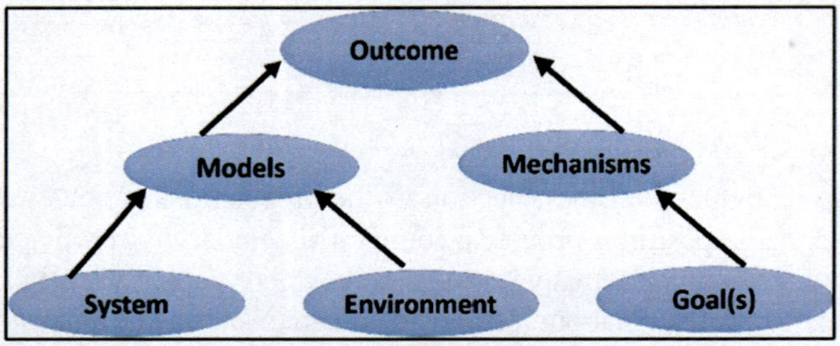

Fig. 1. Proficiency dependency graph for self-assessment

Outcome: The *outcome* vertex represents the evaluation of proficiency, including (a) a binary assertion about whether the agent is proficient or not, (b) an indicator score about the probability that agent can competently solve a problem, or (c) a degree or quality level at which the task can be performed.

Goals, Systems, Environments: The three vertices in the bottom layer represent conditions about the world, system/physical robot, and goal(s) to be solved.

Mechanisms and Models: These two vertices refer to aspects of the algorithms used by the agent to try to achieve its goals given its system and environment. For an intelligent agent, we identify two aspects of algorithms critical to proficiency assessment: *Mechanisms* and *Models*. These are represented as the two vertices in the second level of the graph.

Both models and mechanisms represent assertions about how the agent solves a particular problem. We operationally define a mechanism as a goal specification or set of incentives that explicitly or implicitly encode a goal. Based on our definition, several problem-solving techniques from literature can be thought of as "mechanisms". For example, classifiers often use classification accuracy or precision/recall, MDPs (Markov Decision Process) use rewards, optimal control systems often use objective functions, while planners often use temporal logics to specify a goal.

Models are assumptions made by the agent about (i) how the environment works, (ii) the effect of agent's actions on the environment, and (iii) the relationship between sensors and the environment. It should be noted that the model assumptions are not always explicit, as in reflex agents, and may implicitly be a part of the corresponding mechanism. Examples of explicit mechanisms include: (1) a classifier's use of trees or networks to represent a process by which a decision is made for a given set of training inputs and outputs; (2) an MDP's use of state transition matrices that map present-state-action to next states in a way that represents environment and system; (3) an optimal control system's use of physics-based models for a physical plant, actuators, and sensors; and (4) a planner's use of state transition systems to represent how agent actions affect the world.

Alignment: The mechanism for solving a problem should *align* with the goal, and assumptions about the corresponding model should *align* with the realities of the environment and the system. The connections between the bottom vertices (world, system, goal) and the middle vertices (model and mechanism) represent the two alignment problems, which we refer to as "goal alignment" and "model alignment," respectively.

Consider an MDP problem that can be solved using value iteration, given a set of rewards and a transition probability matrix. Assuming that model alignment holds, an optimal policy may not accomplish a problem holder's goal if the rewards used by the solver do not align with the goal; there is goal-mechanism misalignment. On the other hand, assuming goal alignment, the given transition probability matrix may not correctly model the uncertainties in the system and/or environment, leading to a mismatch between observed distribution of (state, action) pairs and model assumptions. Referring back to Table 1, an agent may not be proficient at a goal (given a system and environment) if either goal or model misalignment exists.

Using Proficiency Alignment: An agent's proficiency at a goal or goals within a number of system configurations and environment is contingent on the conditions of goal and model alignment being met. Misalignments can be useful in generating explanations for proficiency failures because they provide cause and effect reasoning.

For example, suppose that an agent is using an optimal policy derived from an MDP in a long-duration mission. The agent is tracking the empirical history of present-state, action, next state triples. The agent compares the empirical distribution to the transition probability and finds a significant discrepancy. The agent concludes that there is a model misalignment, and reports that its models of the world are likely not sufficiently accurate to perform the mission.

Continuing the example, suppose that the empirical distribution matches the transition probability. Suppose further that the agent has a logic-based recognizer that it uses to determine when tasks are completed. If the recognizer persistently reports failures, the agent can conclude that there is a misalignment between its goals and the rewards used to create the optimal policy. The agent reports that it needs to observe a human performing the task for a while, and then uses inverse reinforcement learning to derive a new set of rewards.

4 Summary and Future Work

This paper presents preliminary ideas for self-assessment of proficiency for an intelligent system. We identified mechanisms and models as the mid-level representative entities, to self-evaluate (i) an intelligent agent's ability of competently satisfying a goal(s) with variations in system configuration and environment settings, and (ii) evaluate cause and effect on (i).

As a part of the future work, we are working on an exhaustive literature review organized according to the definitions of span and the characterization of alignment. This review should provide insight and a narrative into where the state-of-the-art fits within the proficiency dependency graph, what the intended span of the assessment is, and whether the assessments are intended for use a priori, in situ, or a posteriori. Leveraging the findings of the literature review, we plan to formalize our ideas further to develop a generalized framework for proficiency self-assessment of intelligent systems.

Acknowledgments. This work was supported in part by the U.S. Office of Naval Research under Grants N00014-18-1-2503 and N00014-16-1-302. All opinions, findings, conclusions, and recommendations expressed in this paper are those of the author and do not necessarily reflect the views of the Office of Naval Research.

References

1. Daftry, S., Zeng, S., Bagnell, J.A., Hebert, M.: Introspective perception: learning to predict failures in vision systems. In: 2016 IEEE/RSJ International Conference on Intelligent Robots and Systems (IROS), pp. 1743–1750. IEEE (2016)
2. Zhang, P., Wang, J., Farhadi, A., Hebert, M., Parikh, D.: Predicting failures of vision systems. In: Proceedings of the IEEE Conference on Computer Vision and Pattern Recognition, pp. 3566–3573 (2014)
3. Wu, H., Lin, H., Guan, Y., Harada, K., Rojas, J.: Robot introspection with bayesian nonparametric vector autoregressive hidden Markov models. In: 2017 IEEE-RAS 17th International Conference on Humanoid Robotics (Humanoids), pp. 882–888. IEEE (2017)

4. Kaipa, K.N., Kankanhalli-Nagendra, A.S., Gupta, S.K.: Toward estimating task execution confidence for robotic bin-picking applications. In: 2015 AAAI Fall Symposium Series (2015)
5. Israelsen, B., Ahmed, N., Frew, E., Lawrence, D., Argrow, B.: Machine self-confidence in autonomous systems via meta-analysis of decision processes. In: International Conference on Applied Human Factors and Ergonomics, pp. 213–223. Springer, Cham (2019)
6. Lakhal, N.M.B., Adouane, L., Nasri, O., Slama, J.B.H.: Interval-based solutions for reliable and safe navigation of intelligent autonomous vehicles. In: 2019 12th International Workshop on Robot Motion and Control (RoMoCo), pp. 124–130. IEEE (2019)
7. Havens, A., Jiang, Z., Sarkar, S.: Online robust policy learning in the presence of unknown adversaries. In: Advances in Neural Information Processing Systems, pp. 9916–9926 (2018)

Behavioral-Based Autonomous Robot Operation Under Robot-Central Base Loss of Communication

Antoni Grau[1(✉)], Edmundo Guerrra[1], Yolanda Bolea[1], and Rodrigo Munguia[2]

[1] Automatic Control Department, Technical University of Catalonia UPC, Barcelona, Spain
{antoni.grau,edmundo.guerra,yolanda.bolea}@upc.edu
[2] Computer Science Department, CUCEI, University of Guadalajara, Guadalajara, Mexico
rodrigo.munguia@academicos.udg.mx

Abstract. Robot navigation requires the use of a reliable map. Depending on the environment conditions, this map needs a constant update for a safe navigation. Autonomous robots use this map but at the same can contribute to the updating process, requiring a permanent connection to the cloud where the map is created and modified based on the robots' information. In this paper, authors present a robot navigation scheme based in hybrid control behavior when the connection to the cloud is loss for some reason. Robot needs to recover a known position to restart its mission and the behavior definitions allow this fact. Results with real data are presented in front of different situations of network coverage.

Keywords: Autonomous systems · Robotic agents · Behavioral control

1 Introduction

In the last years, autonomous robot navigation has become a great field in robotics. Many applications and commercials products rely in autonomous navigation (Google car, Tesla car, GM Cruise… [1–3]), and it looks robot navigation as a solved fact. In reality it is not clear that navigate with an unmanned vehicle has a direct solution. Apart from navigation, the creation of the map is an open question in robotics research. The most popular technique is SLAM (Simultaneous Localization and Mapping), and many efforts have been devoted to this set of techniques in the last years [4, 5].

Most of the works that one can find in literature are based in the creation of a map using a single robot, which explores the environment, creating a map, navigating in this map and locating itself in the map. This sequence of actions requires huge load of computation, which is not always available in autonomous systems that are built with embedded processors with a reduced power of computation. An alternative is to use ubiquitous computing through the cloud. The concept is clear: the robot sends environmental information, sensorial information and the current map to a computer (or set of computers) that are in the cloud, then a new map is computed based on this uploaded information and the robot receives a new and increased map to locate itself. The process

M. Zallio (Ed.): AHFE 2020, AISC 1210, pp. 114–120, 2021.
https://doi.org/10.1007/978-3-030-51758-8_16

is repeated until the entire environment is explored and the full map is created. Another alternative is the creation of a collaborative map through a fleet of robots connected to the cloud and sending parts of their environment in order to create a more general map fusing and combining all the small and partial maps created and explored by every autonomous robot [6, 7]. This technique is difficult because disconnected map are created and need to be glued at every step of the process.

In both cases, the use of cloud computing is a common point that are shared in the alternatives. When the global map is built therefore it can be shared (or let) between robots to navigate in the already explored environment. Because some of the environments can change for unexpected reason (works, risky areas, rescue areas after disasters…) the map to navigate is updated all the time in a continuous SLAM process. In most of the cases, this procedure requires permanent connection to the processing unit. The unit in charge to process all the information sent by the robot will be in the cloud. Therefore, a permanent connection to the cloud is needed to maintain the map updated.

Robot control is a concept that can also be defined as robot decision-making or robot computational architecture. It is defined as the process of acquiring information about the environment by the robot's sensors, processing this information and to make decisions about how to act, and executing actions in the environment. The complexity of the environment has a direct impact on the complexity of control, which is, in turn, directly related to the robot's task. There are mainly four classes of robot control methods:

- Deliberative: Think, then Act. In deliberative control, the robot uses all of the available sensory information, and all of the internally stored knowledge, to reason about what actions to take next, [8].
- Reactive: Don't Think, (Re)Act. Reactive control is a technique for tightly coupling sensory inputs and effector outputs, typically involving no intervening reasoning to allow the robot to respond very quickly to changing and unstructured environments [9].
- Hybrid: Think and Act Concurrently. Hybrid control aims to combine the best aspects of reactive and deliberative control: the real-time response of reactivity and the rationality and optimality of deliberation [10].
- Behavior-Based Control: Think the Way You Act: Behavior-based control employs a set of distributed, interacting modules, called behaviors that collectively achieve the desired system-level behavior, [11].

In this paper, authors propose a navigation procedure based in robot behaviors with the objective to reach a certain destination marked in the map. The different situations that the robot can encounter through the pathway have to be analyzed and a behavior has to be defined depending on each situation. Specifically, in this paper authors propose a behavior when the connection to the cloud coverage is lost. The behavior will be the alternative to stop the robot and await to recover the signal which is unclear that could be recovered. Robot has to work with no a priori knowledge about the reason why the network coverage is lost.

The behavior has been implemented in a real robot, which is programmed to reach a specific destination in a real environment. In Sect. 2, the methodology used to define and implement the behavior is shown, exploring different behaviors in front of several situations that robot can find in the environment. Section 3 describes the equipment and the environment used to test the robot behavior in front a loss of coverage during the pathway to its destination, with some interesting in real situations. Section 4 concludes the paper together with the references.

2 Methodology

In this paper, authors propose the use of a hybrid control architecture that has proved to work with good results in dynamic and complex environments. In the next section, the autonomous robot that has been used to testing the architecture is explained.

The propose architecture can be seen in Fig. 1, depicting the integrated hybrid navigation system for autonomous robot.

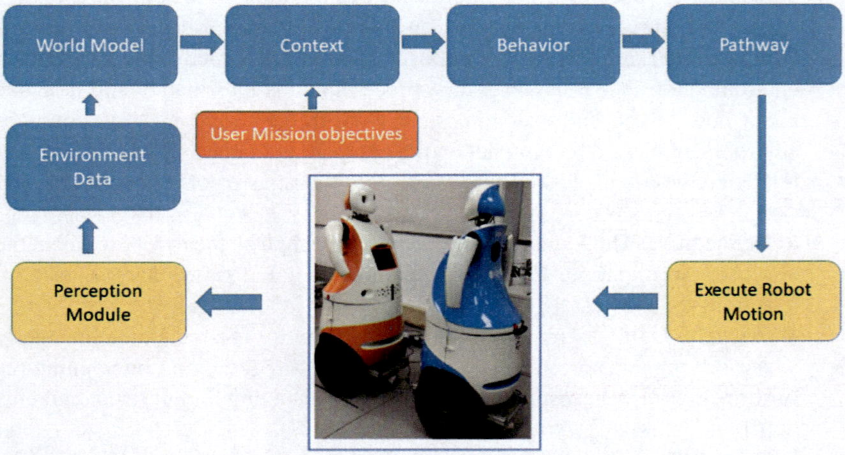

Fig. 1. Hybrid navigation system for a mobile robot architecture.

The world model is in fact the robot's environment, represented in some different layers where each one represents a characteristic of the environment. The robot state is represented in one of those layers with its pose (location and orientation). The world model is built using two kinds of data, local and global. Local data is obtained with intrinsic robot sensors (odometry, range, vision, Wi-Fi connection). This data will be used to navigate without colliding and to locate in the map. Global data is obtained from the cloud and mainly contains the updated map of the robot environment. This data will be used to plan the pathway between origin, waypoints and destination of missions.

With all this information, robot can obtain its context and plan a navigation pathway. Context also contains the set of available strategies to execute the mission that the user specifies. The different robot kinematic and dynamic model are also defined in the context as well as the state of navigation (either reactive or deliberative). No obstacles are preset because they can appear dynamically/suddenly in the environment and global map does not contain them.

The behavior block contains the deliberative and reactive behavior implemented in the robot. As a feature in this block, behaviors can be download from the cloud depending on the evolution of the robot, its environment and its mission. This is another important constraint to maintain permanent Wi-Fi coverage in the best of the cases. Based on all the previous blocks, the pathway is planned, and robot starts navigation, sensing continuously local data and uploading/downloading global data.

The behaviors that authors want to emphasize in this paper are i) *normal navigation with Wi Fi coverage* and ii) *reduced navigation without Wi-Fi coverage*, and the differences between them.

2.1 Normal Navigation with Wi Fi Coverage Behavior

If the mission of the robot is to go from an origin (A) location to a destination (B) location, the deliberative behavior is obviously to plan a pathway to reach B point in the map. The procedure to do this action is based in a graph description of the environment (extracting keypoints based on range sensors and computer vision) and it is explained in a previous authors' work, [12]. The reactive behaviors are checking for two events: obstacles in the pathway and keeping Wi-Fi coverage during the way to exchange data. If none of these events happens, the robot will follow the planned pathway based on local data obtained from sensors and navigating with the downloaded map.

2.2 Reduced Navigation Without Wi-Fi Coverage Behavior

The mission is always the same, to navigate from points A to B in the global map, and therefore the deliberative behavior is the same than the previous situation (Subsect. 2.1). However, when the Wi-Fi signal sensing function detects a loss of connection a function (as an exception) is executed, see Algorithm 1.

Algorithm 1. Function to treat the loss of connection to the cloud try to recover it in another location of the environment

```
function no_signal_detected(path_type *path) int
{
 start_timer(&timer);
 while !signal(wifi) && timer < maxtime_loss  timer=read_timer(); end
 if signal(wifi) return(1)
  else                       /*permanent loss of connection in location*/
     robot_motion(recover_coverage_situation);
/* stop, turning 180 degrees and go forward at slow speed until recover
network coverage again*/
     *path = recalculate_pathway(actual_position, pointB);
     if *path == NULL     /*no pathway connects A and B*/
          *path = recalculate_pathway(actual_position, pointA);
     end
 end
return(0)}
```

When signal is lost, the function in Algorithm 1 is executed. During certain time (`maxtime_loss`) the robot keeps on running through the pathway because instantaneously losses of connection are expected (see results' Section), and it is possible that network coverage is recovered. If this threshold time is passed, then it is considered that the coverage is permanently lost in the robot's location and a new pathway has to be calculated from the present location to the final destination (point B). To recover coverage again, the robot receives the instruction of stopping, turning 180° and advancing forward at low speed until the network is recovered again, at that very moment, the new shortest path is calculated again. If this new pathway has permanent coverage then the mission will be completed, but if any of the possible pathways cannot connect A and B then the behavior is to return to point A, the origin, where there was network coverage because robot departed from that point. Obviously, there is emergency behavior to avoid that robot goes turning infinitely if coverage is lost due to a failure of the network equipment or similar. This behavior is not indicated in Algorithm 1.

3 Results

The robot to be used to perform the experiments is based in a Segway platform equipped with GPS (unable to work under certain circumstances, GPS-denied environments), cameras and range sensors. Robot is equipped with a Wi-Fi equipment in charge of connecting the robot with the cloud. Experiments have been carried out in Barcelona Robotic Urban Lab, a tagged European robot facility located in BarcelonaTech campus [13] (Fig. 2).

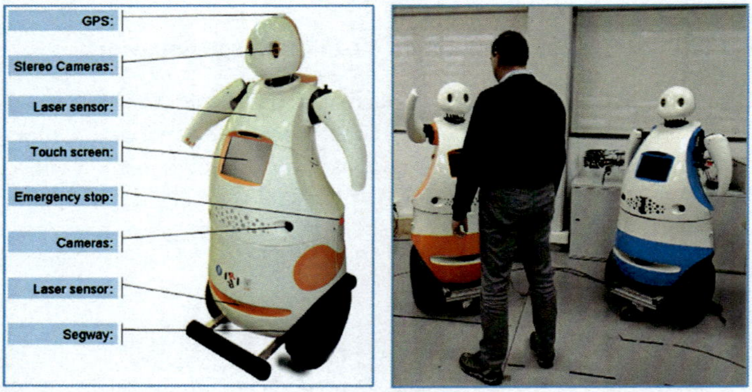

Fig. 2. Robot used for real experiments and actual size.

The above behavioral architecture has been implemented in the robot, and we have given access to different Wi-Fi router, depending on the situation we want to evaluate. In Fig. 3 a), there exist network coverage in the shortest path between A and B. Therefore, the system is in the situation explained in Sect. 2.1, with normal execution of the navigation pathway.

In Fig. 3 b), the situation changes and we have disconnected one router leaving an area in the shortest path without network coverage, and then robot enters in behavior of reduced navigation. When robot enters in the no-coverage area, after a specific time, robot stops, turning and entering again in the coverage area, afterwards robot follows the recalculated pathway avoiding this area. Again, robot enters in a uncovered area but

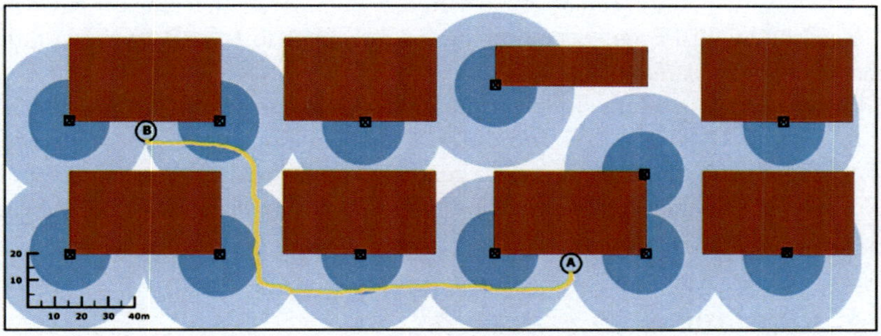

a) First scenario: full coverage along the pathway between A and B locations.

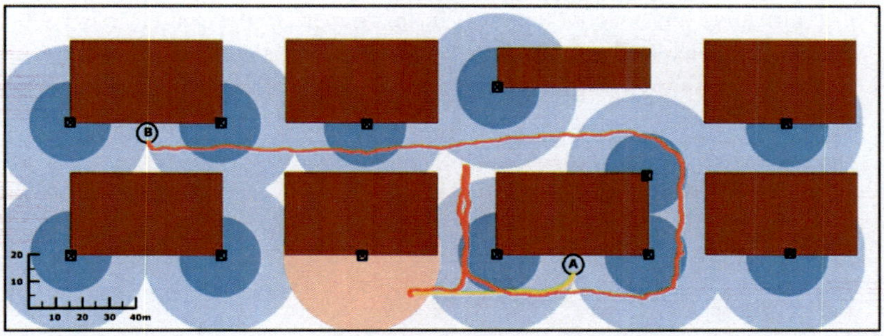

b) Second scenario: lack of network coverage in one area but the destination has a feasible pathway with network coverage.

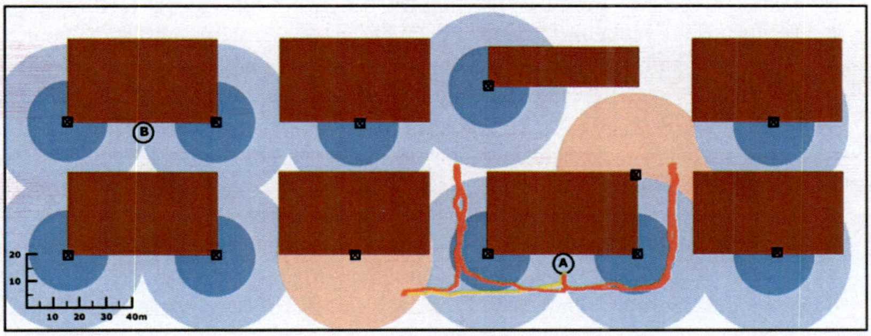

c) Third scenario: there is no network coverage between origin and destination.

Fig. 3. Three scenarios with different situation of Wi-Fi coverage. Blocks in brown correspond to the Campus buildings, and blue circles are the coverage given by Wi-Fi router located in black spots in building outer walls. Scale allows seeing distances and network coverage radius (only outdoor is shown) for the routers.

its time is over under this situation and executes the same procedure (stop and turning), searching for a new pathway to the destination B. This is the yellow path plus the red path showing both status after executing the behaviors.

The last situation can be encountered in Fig. 3 c). In this case we have disconnected a couple of Wi-Fi router, cutting any path from A to B, and after two failed tries for reaching the destination, no available path exists between location A and location B, for this reason robot returns to the origin point A, awaiting for a new mission.

This test has been executed different runs, giving always the same results in terms of the mission, obviously, the path generated by the robot is slightly different from one run to another and even the traveling time is different, but those aspects are matters that have no interest in this research.

4 Conclusions

Hybrid behavior control has been proved to yield excellent results, with a combination of deliberative and reactive control. In this paper, an experiment has been carried out with such a behavioral architecture with a robot in a real environment. The robot can relocate its position though the network coverage is lost during the use of a map in the cloud.

Acknowledgments. Spanish Ministry of Economy, Industry and Competitiveness through Project DPI2016-78957-R funded this research.

References

1. Autonomous Google Car: Waymo. https://waymo.com/. Accessed 28 Jan 2020
2. Autonomous Tesla Car. https://www.tesla.com/autopilot. Accessed 28 Jan 2020
3. Autonomous GM Cruise Car. https://www.getcruise.com/. Accessed 28 Jan 2020
4. Trujillo, J.-C., Munguia, R., Guerra, E., Grau, A.: Visual-based SLAM configurations for cooperative multi-UAV systems with a lead agent: an observability-based approach. Sensors **12**(4243), 1–30 (2018). https://doi.org/10.3390/s18124243
5. Yang, T., Li, P., Zhang, H., Li, J., Li, Z.: Monocular vision SLAM-based UAV Autonomous landing in emergencies and unknown environments. Electronics **7**(73), 1–18 (2018)
6. Michael, N., Shen, S., Mohta, K., et al.: Collaborative mapping of an earthquake-damaged building via ground and aerial robots. J. Field Robot. **29**, 832–841 (2012)
7. Minaeian, S., Liu, J., Son, Y.J.: Vision-based target detection and localization via a team of cooperative UAV and UGVs. IEEE Trans. Syst. Man Cybern. **46**, 1005–1016 (2016)
8. Albus, J.S.: Outline for a theory of intelligence. IEEE Trans. Syst. Man Cybern. **21**, 473–509 (1991)
9. Shusong, X., Jiefeng, H.: Biologically inspired robot behavior design. In: 6th IEEE International Conference on Industrial Informatics, pp. 1–6. IEEE Press, New York (2008)
10. Malcolm, C., Smithers, T.: Symbol grounding via a hybrid architecture in an autonomous assembly system. In: Maes, P. (ed.) Designing Autonomous Agents: Theory and Practice from Biology to Engineering and Back, pp. 123–144. MIT Press, Bradford Book, Cambridge (1990)
11. Mataric, M.J.: Reinforcement learning in the multirobot domain. Auton. Robot. **4**(1), 73–83 (1997)
12. Guerra, E., Munguia, R., Bolea, Y., Grau, A.: Human collaborative localization and mapping in indoor environments with non-continuous stereo. Sensors **16**(3), 275 (2016)
13. BRUL: Barcelona Robotic Urban Lab. https://www.iri.upc.edu/project/show/213. Accessed 28 Jan 2020

Author Index

M. Zallio (Ed.): AHFE 2020, AISC 1210, pp. 121–122, 2021.
https://doi.org/10.1007/978-3-030-51758-8